IDEASTHESIA

Also by Vera Dragilyova

IDEASTHESIA

Artificial Intelligence Inside the Human Brain

Vera Dragilyova

Verarta Books

CONTENTS

PART IV—VISUALIZATION OF THE WORLD

INTRODUCTION

This book is a first person account of an alternative way of processing reality, venturing into an area of neuroscience that is still largely unexplored. The focus of the book is on Ideasthesia, a neuropsychological phenomenon where one experiences abstract thought in physical ways, through the five human senses. Synesthesia, where sensory perceptions in one sense are accompanied by perceptions in one or more other senses, is also addressed. It is still unclear to science how and whether the two are related.

The title of this book was inspired by the visual resemblance between the diagrams seen in the Artificial Intelligence textbooks and the processes in the brain, as experienced by the author through Ideasthesia.

Much of the research in Artificial Intelligence is an endeavor to model itself after the human brain. Yet, understanding human brain's organization, as well as consciousness, still remains an elusive goal, despite our ability to scan and dissect the human brain, and even observe neurons under powerful microscopes. Being a unique visualization tool, Ideasthesia might provide a peek into the workings of the brain that is otherwise unavailable to science.

The book presents an array of topics, as they are experienced through Ideasthesia, and does not delve deeply into any one of them in particular. It is written with a hope that such a foray deep into self will become useful to science, if not for exposing something yet undiscovered, then for posing engaging questions, and offering a completely new perspective on the workings of the human brain.

The author extends an open invitation to everyone to discuss, question, and test her ideas, as only through tireless trial and error can one find a way out of a labyrinth, and only through chiseling away the extra pieces can one reveal the angel inside.

Welcome to my world!

THE BEGINNING OF THE BEGINNING

You can't make me work inside an office, just like you can't staple water.

One fine afternoon, at a nice sunny cafe, my good friend Elsa was sitting across from me at a small table. Sun, sprinkling us with sunshine, there was a whole while of silence to take it all in. Motley shadows of trees shimmering on the surface of our wooden table, its aged cracks made me think of all the people who sat here through the years, pouring out their thoughts, hopes, and secrets to someone. I closed my eyes and slid my finger across the rough surface, as if hoping to trace what was shared with the table by its solitary sitters, as the ball of their pen etched their writing deep into the paper, inevitably imprinting it on the wood. This is how blind people read. But what if someone was only typing? Would the air around remember every movement of their fingers?

"Elsa, if I don't finish this book, I will prove to myself that life amounts to nothing."

"Vera, how old are you?"

"I am five at home, fifteen in the public."

"And otherwise?"—exclaimed Elsa with marked horror that dug a deep wrinkle between her eyebrows, resembling the table's cracks. Despite our huge age difference—Elsa was some decades older than me—we always found a common language.

"Otherwise, I could be a grandma like you. But, if I do write this book—why would that even matter?"

"Oh, dear, I think, that's quite impossible."

"Why? I specialize in the impossible, Elsa!"

"So what is all the raucous about?"

"I can't think and write at the same time."

"Well, you could tell me your story and I could take notes. Then, you could use these notes to write the book yourself. Can you think and speak at the same time, at least?"—she laughed.

"Think deeply—no, the sound of my voice erases all thoughts in my head."

"Oh… What if you just talk without thinking?"

"Perhaps, that's the only solution. That means bypassing the logic filter of the conscious, and speaking straight from the heart. The brain, that is, because that's where the heart really is. Thinking is my favorite sport, Elsa, you know that, but…"—I answered, slowly tracing the wooden crevices, and with a quivering voice.

"But what?"—she asked with barely suppressed impatience.

"Elsa, I would feel so incapable. Think about it: Elsa—my angel Gabriel in the flesh!"

"Then, it's a yes?"—joyfully asked Elsa, with a big smile from ear to ear.

"I am just scared! What if no one will understand?"

"You know, whatever you say can be used against you in the court of law,"—said Elsa, savoring her every word.

"When I get your age, I want to be like you—not worried about anything, as if there is nothing to lose,"—I could feel my lips pulling into a trumpety pout.

"Well then! If you don't lend your thoughts to air, they will never land on my ears, and certainly—not on a single sheet of this glistening white paper. Tell me your secrets, Vera, and I might even tell you mine."

"I already know all your secrets,"—I grimaced disappointedly.

"It only seems so because you are young. Just start with a little secret, and it will snowball from there!"

"My big secret is that I see and feel what I think."

"What?"

"Ah, you must be old school. So, I feel my ideas just like I feel this hand of mine, right here,"—and I stretched out my palm all the way up to Elsa's nose.

"Not too close! I am not a microscope."

"Look deeper, Elsa."

"I don't need to look, I only need to listen to the big picture—so that I can write it all down. Here is how it is going to work: since I am older, I will be the captain of the ship. So, now, give me something livable and writable."

"It is hard. I have never thought in words, I have always thought in chemical reactions, physical forms coming alive in my head… I am them, and they are me. In this phantom reality of mine, I could always become any object and feel its pain, and the world has always felt like it was a part of my body, and I could transport myself anywhere in time or space instantaneously…"

"Vera, stop! You are such an obscurantrisse! Just give me a straight story."

"Behind every story in the world, there is a world of curly secrets, Elsa, you know that. This secret truth inside me—I have no other story to tell."

"I can write down only what I understand, Vera."

"But if you get mad at me, I will get mad at you for getting mad, ok?"

"I am not what I seem, Vera."

"Elsa, you've got me by my brains!"

"The most vulnerable part of your body,"— Elsa chuckled. "Start thinking, so I can help your life amount to something."

"One more thing: there is another problem. Whenever I wake up the next morning, I forget everything, good and bad."

"That means, if you get mad at me for pushing you too hard, you will forget that, too. Or maybe you will even forget about me?"

"That only means that we have very little time —only until the sunset."

MY STORY

Elsa looked straight into my eyes, with a squint of a grandma who already knows everything about you, but only asks questions to catch you on a lie: "So, how did it all begin?"

"I have been like this since as far as I can remember. But then, this one time… No, that's not important."

"Tell me, what happened?"

"Once on the phone, I was explaining my summer plans to my mom. It had to do with graduate school, and where I was traveling, and what I still wanted to learn. When I finished my explanation, I only heard a long pause of silence. "Wait, you are all over the place, I don't understand anything!"—my mom finally said, with sandpaper frustration.

In my head, it was all crystal clear. I saw a vast terrain beneath me, as if I were hanging over it, stretching left to right—more exactly, west to east, and also north to south, if I were to rotate it. The terrain was like a skeleton with fillable intertwined pockets. There, spread out, were all my activities in the form of objects: solid were the ones with less flexibility of getting accomplished, and soft and stretchable ones were those that could be stretched out in time, or paused if necessary. I knew which one had to be done first, so that another one had a chance of happening, as if in a chain reaction. It looked like a project management board, with all the month and year borders clearly marked—months varying in color, depending on my emotional relation to them. I could hover and zoom into any month, day, hour, minute, or second, like a humming bird into a flower. Altogether, it felt like an emerging complex system that I could model by trial and error, as if through a mass of unfolding Chaos theory of that virtual space, and without risking anything in real life.

I thought that my mom had the same experience, so I asked her about it. She said, "Nothing of the sort, this sounds like total chaos!"

"Not the chaos you are imagining!"—I thought to myself. "You really don't see what I see the next summer?"—and I waved my hands in the air, as if she could see me over the phone. "It is all right here!"

By that time, I was already used to being misunderstood, and having to explain something that seemed so simple to me, in other 'boring' ways. Yet, somehow, this very moment, it dawned on me: people *don't* see what I see, as I hoped all these years. So, of course, when I refer to something in my visualized reality, they have no clue as to what I am talking about! And I used to think that I was simply inarticulate, or we disagreed about what we were seeing, whereas the question I should have been asking is whether they saw anything at all. Why didn't I ever ask anyone whether they actually *saw* and *felt* the next summer and what it looked like to them? I never asked anyone *where* next Tuesday was, or how the longing for summer will weigh by the time the summer comes, for example. I just assumed that they saw what I saw. When I was little, I did ask people how they did multiplication in their heads, what they saw when they did it, and never really got a satisfying answer. So, something along the way made me close in, and started me thinking that everyone else was better than me.

I thought of my days in an American High School, where I started as a fresh-off-the-boat Soviet immigrant. The first thing I heard at school was "you are weeeeird", which came from a red-haired girl who was sitting in front of me, in the first class of the first day of school. I only asked her a simple question, but she could not understand, and, as she paused, she squinted and fixed her eyes on me. Her blond eyelashes bunched up together, as she whispered right into my face: "You are weird!", and turned around. I did not say anything, but almost said thank you, as a default. Just because I already knew that I was weird. In the United States, as I had just discovered, being different was a good thing.

In college, when my friends heard that my major was communication, they joked that it was really *miscommunication*, because I always managed to twist things into mentally un-chewable ideas that nobody seemed to swallow. Although, these ideas made perfect sense to me, so I would call them quite *chewable*, and would have chewed on them anytime, if you asked me. And on top of things—my ideas worked! My ways of thinking got me to understand complex topics very quickly, and predict the outcomes correctly, although I had nothing to show for how I got there.

For some odd reason, this conversation with my mom was a turning point in my self-awareness, even though there was nothing drastically new about it. It wasn't the first time I got this reaction, yet this time, it became a firing trigger to figure things out, once and for all.

This world in my head could not possibly be chaos and insanity, because it clearly was full of sense, beautiful, elegant, and even fantastic, and I just knew that I wasn't crazy. There had to be something else. Something that nobody knew about yet. This something would explain why in college, I skipped from major to major, and ended up graduating in Interdisciplinary Studies—because I obsessively wanted to learn it all. Also, why I have lists, and list of lists, of things I want to learn, do, visit, taste, read, experience, and my life dreams always come in complicated knots. This would explain why my brain is always looking for completion of a puzzle, and every time, when it completes it, the answers it gets are confirmed as right. A big picture puzzle needs to be completed in my internal reality, so that it can study it, model it, and have a grasp on its entirety. Otherwise, it would hurt and bother me non-stop.

"I don't see anything!"—all exasperated, said my mom.

"Then, how do you think?"—I asked with equal exasperation.

"I don't know, it just comes to me,"—she answered.

"I cannot even imagine what that must feel like, a dark void? Death?"—I thought to myself, but I only said: "Ok."

At this very moment, I decided to write a book about this, and have as many people as possible read it. Maybe somebody, someday would understand me."

"We can do this! Yes, we can!"—exclaimed Elsa with unexpected joy, loudly slapping her hands on the table, so it shook and almost fell apart. "Tell me more! And… so? Then what?"

WHAT SYNESTHESIA

FEELS LIKE

"I am afraid, Elsa."

"What is your greatest fear, Vera?"

"It is saying nothing new."

"After everything you have said, I wouldn't be too afraid,"—she chuckled. "Have you done any research on this at all?"

"I have not done any research, except... I think that my way of thinking is called Ideasthesia. This is when a person experiences abstract thought through the five senses: touch, vision, smell, haptics... I also have all kinds of Synesthesia, which is experiencing input from one sensory channel through another. Like seeing music that you hear."

"Should we begin with Ideasthesia?"

"Synesthesia should go first, just because more people know about it, and because it might better to explain a few things, before we get to burrowing through Ideasthesia's jungles. I think, we all have some sort of Synesthesia. The only difference is our awareness of it, so let me tell you about this first."

"Sure, just make sure you stay clear. Because, ya' know, I don't have those visions of yours!"—she chuckled her repetitive chuckle that, I felt, was starting to get under my skin, and I thought to myself: "Why is she so condescending?" My eyelids felt like two water dams pregnant with explosives. "Synesthesia is at the root of anthropomorphism of animals and inanimate objects, seeing special significance in random events, and reaches as far as superstition and even ability to conceptualize religion." Elsa laughed her lungs out.

"Ha ha ha! Well, that is a good start!"

"I don't know another way to put it, but…"

"The proof is in the putting, Vera, so tell me more!"

"So, for example, numbers all smell good, especially the red ones."

"Hm... there is something wrong with this courtroom."

"Oh, that again? Am I now a defendant?"

"Don't tell me you didn't know that."

"You must be the judge. Then, your Honor, please honor me with hearing I will tell you about all kinds of Synesthesia that I have."

"If the Letter of the Law allows so, I will."

"Let's start with numbers, and follow with letters. My numbers all have colors and personalities, temperature, weight, and who knows what else."

"You are the only one who knows, Vera."

"Here they are!"

Numbers

0

COLOR: Clear with thin black frame that takes the shape of whatever it is introduced into— just like water.

PERSONALITY: I would not say that it is kind. Neither man or a woman.

TEMPERATURE: The same as my skin.

WEIGHT: Weightless.

SMELL: No smell.

TASTE: Tastes like water, but depending on the context of thought, it can taste bitter or light sweat. Never-ever salty.

1

COLOR: Pure white.

PERSONALITY: Inspirational, strong, forgiving. Both masculine and feminine.

TEMPERATURE: Cold, but pleasant.

WEIGHT: Immovable, you cannot seem to move it or pick it up.

SMELL: Since I see smells, it looks grainy. It has a powdery smell.

TASTE: It can taste salty, but usually doesn't have much smell.

2

COLOR: Warm bright red.

PERSONALITY: Brilliant, courageous, proud, a leader, can be rebellious. Definitely feminine.

TEMPERATURE: Hot and burning.

WEIGHT: It is pretty heavy, but liftable.

SMELL: It smells sweet, probably because it is red, but not overpowering.

TASTE: Sweet, but not overpowering.

3

COLOR: Lemon yellow.

PERSONALITY: Kind, agreeable, I could have it as a friend. Feminine.

TEMPERATURE: Cool, but pleasant.

WEIGHT: Light and feels to hold.

SMELL: Sweet fresh smell, I love it!

TASTE: Light sweet and sour, aromatic, fresh!

4

COLOR: Pitch black.

PERSONALITY: Stoic, mysterious, dangerous. I am afraid of a four! Sharp angles. Could either be man or a woman.

TEMPERATURE: Has no temperature, because it would not tell it to me.

WEIGHT: Heavy, unmovable, unreachable.

SMELL: Metallic smell, for sure!

TASTE: Metallic, maybe even salty and bitter. Very unpleasant.

5

COLOR: True red.

PERSONALITY: Kind, strong, happy. Definitely a woman.

TEMPERATURE: On the hot side, but pleasantly so.

WEIGHT: A bit heavy, but easy to pick up and throw.

SMELL: Sweet, flowery. It is a female!

TASTE: Sweet and pleasant.   This is my favorite number!

6

COLOR: Warm yellow.
PERSONALITY: Timid, quiet, a bit boring. Definitely a female.
TEMPERATURE: The same as that of my skin, and never warm or hot.
WEIGHT: Light.
SMELL: Sweet and fresh, but unpleasant
TASTE: Lightly sweet and watery.

7

COLOR: Heavy yellow.
PERSONALITY: A dominant female.   Not rebellious like 2 or prominent like 5, though, and older than them.
TEMPERATURE: On the warmer side.
WEIGHT: Heavy, heavier than 5.
SMELL: Like heavy perfume for old ladies, although it has fresh notes, too.

TASTE: I don't like it! Like nothing that humans eat. Grainy and powdery.

8

COLOR: Bluish, almost purplish red, maybe fuchsia?

PERSONALITY: Mysterious, kind and smiling female.

TEMPERATURE: Cool.

WEIGHT: Light and elusive.

SMELL: Light mysterious smell, smells like the night.

TASTE: There is something pleasantly sour in it, but very light.

9

COLOR: Medium green.

PERSONALITY: A kind, confident male. Smart.

TEMPERATURE: Neutral, but definitely not cold.

WEIGHT: Not too heavy, but has weight.

SMELL: Smells like a lighter man's perfume.

TASTE: Salty, for some strange reason.

"What's funny is that I do not have any numbers expressly in blue. 5 and 8 have some blue in them, but none of them is all blue. I wonder why... When numbers come together, I see them as people standing together, and some of them don't get along. For, example, everyone is afraid of the zero (0). Four (4) does not like anyone, and everyone is afraid of it. Two (2) always wants to win over everyone else. Everybody likes five (5), because she is pretty, friendly, and confident. Seven (7) is also beautiful, but she is quieter and less outspoken than five (5). All the females like number nine (9). If there is a number 10 or, say, 345, they chemically blend together and lose their personalities. I feel like they really suffer from it, as if they are tied up in punishment from Gods, when they get into those clusters! I have something else to say. It has to do with Mathematics. Guess what? Multiplication is sticky, addition is dry. When equations are not solved, they feel like imbalanced knots, waiting to be unraveled. Simple two variable equations feel like onions to be peeled, to get to the center, where the secret unknown resides. All equations feel like living creatures, where what comes in, must come out, after it is digested! Input—output, call—response.   Or a broken phone, where the equation is what garbles the message. Some number combinations, formulas, and equations are pretty and some are extremely ugly.   I love numbers that feel smooth—those that end on zeros. Or those in red and yellow."

"Ok, interesting. Next?"

"Letters. By the way, I cannot stand typing in a sans serif font, because I feel like my eyes cannot grab onto them, and keep slipping. Those letters are so oily! I need those little serifs—the angels!—to grab onto. I feel like I have tiny visual hooks holding onto each one of them, and it feels really comforting. Otherwise, it is like I am walking on water!"

"Write me a letter about letters, Vera!"—laughed Elsa out loud.

"Anyway… They all have colors that describe their personalities.   All letters have a body, complete with face, torso, and limbs. I see which direction they are facing, and whether they are smiling, and whether they are looking at me: all their non-verbal behavior. Also, if they are in a good physical shape."

"If they are so human, why don't you give them names?"

"Because they are like people to me, and I am a people person! I will remember people's faces and their stories, but their names..? Not really. Although, I do tell people: 'You are definitely not a Rebecca! You are such an Angelina! Or, why not name your baby a Grok?' We all know how that feels, right? That's another clue that we all might have Ideasthesia, and that's exciting!"

"And I am excited to hear about each letter."

"Of course, this is only about the letters in the English alphabet. Other writing systems have their own societies of letters, with their own social dynamics."

Letters

a—bright red, a confident female.

b—blue, a kind male.

c—yellow, a blond male.

d—dark green, a dark-haired male, or yellow—a pregnant and happy female.

e—light green, a young friendly male.

f—dark blue/black, a mysterious male or female.

g—dark brown, an old grandma.

h—medium blue/green, a young boisterous male.

i—medium light blue, a very young male, cute.

j—grey, older rich male with good manners.

k—black, a tomboy, or an eccentric young male, or someone with tattoos

l—white, an agreeable female or male with no backbone, one who cannot say no

m—medium dark blue, a strong female or a kind male with a stocky body and muscles.

n—medium green, a girl with particular facial features that I know, but cannot explain, nice and friendly, but quiet.

o—white, male or female, somehow I don't like them.

p—dark brown, Devil's Advocate, surly and grumpy, old.

q—grey, full of ideas, energetic, mentally sharp, quick thinker, male or female.

r—red, energetic, positive, a leader, a female.

s—red violet fuchsia, most likely a female. Weird personality and not trustworthy.

t—black, principled male, trustworthy, straightforward.

u—dark blue, pensive, grounded male.

v—dark pink, definitely a female, always in a good mood, and strong.

w—blue/grey, does not have much to say, a slow male or an old female.

x—black, a male, strong, smart, but uncompromising and can be problematic.

y—very dark blue/green, a nice male, young, friendly, supportive.

z—mahogany, gets along with x, an eccentric female, but friendly and adventurous.

"Playing with fonts is literally a brain massage, by about as many hands as there are letters. It is like coming to a huge party and talking to all the people at the same time, without getting confused or overwhelmed. An endorphinal bath to swim in! However, letters can get stifled and petrified when they link together in words: they suffer from those bonds. They lose their personality, you could say that they die!"

"Everything must come to an end one day. Even letters. What is your favorite letter?"— asked Elsa with a malicious squint.

"I like the letter 'A'!"

"I thought so!"—answered Elsa with an invasive smile. "Do you see yourself as that letter?"

"I think that I would like to be like an A, and also like the number 5, and I wouldn't mind them as friends. They are both bright red."

"Do you see all those numbers, personalities and different qualities on the page when you look at numbers and letters?"—perplexedly asked Elsa, without releasing either her squint or the smile.

"No, I don't see them on the page. I see them only in my head, my Phantom Reality."

"What is that, and where is that exactly?"

"It is in the same place where you would find your memories and your dreams about the future. And all your imagination—in your mind's eye!"

Elsa searched around with her eyes, stochastically looking into space, focusing on nothing that was visible to anyone in front of her. I asked her: "See, right now: where are you looking? It is not on anything that is in front of you, right?"

"Maybe my eyes are searching within my brain."

"Your Phantom Reality?"

"Maybe everyone has one. They are just unaware."

"I am sure of it as I am sure of your helping me with this book!"

"That's beyond a possibility."

"A probability beyond a 100%! That's a beautiful concept. I need to feel it through. It might mean that you have no other choice—because anything other than that— is impossible!"

"I don't specialize in the impossible like you. But, that would make it an infinite probability, wouldn't it?"

"I have no idea! I feel like for a second, we switched places!"

"The more I listen to you, the more convinced I become that I have Synesthesia, just like you. You are right, everybody has it. Otherwise, where would they be looking, if not in their Phantom Reality, when thinking with their eyes?"

"I shall not disturb that thought in you,"— I said with coyness and very subtle note of happy irony in my voice. I like it when Elsa is content! I continued with a big smile on my face: "Days of the week might be also interesting."

Days of the Week in Color

Monday—blue
Tuesday—green
Wednesday—yellow
Thursday—black
Friday—red
Saturday—orange
Sunday—white

"It is funny how days of the week are a little like numbers—you can tell where they are located by their names: Tuesday is always after a Monday and before Wednesday. It is the same with the months. Human names—they are all over the place, non-committal, come and go, as if this is a drive through!"

"Ah, you have a blue day—Monday! Does that mean, you get blue, like everyone who doesn't want to go back to work?"—asked Elsa, as if she did not hear a word of what I just had said.

"Not at all! Blue does not even hint on sadness for me!"

"It sure does for everyone else!"—Elsa answered grimly.

"And, anyway, I did not purposely assign colors, personalities, or anything else, for that matter, to any of this stuff. That is all part of who they are. All I do is describe them."

"Do you have any idea why your days of the week are colored the way they are?"

"No, not at all. I thought, maybe it had something to do with the first letter in their names, but that does not explain it. Maybe they are just alive."

"Do they have any other qualities, besides colors?"

"Not really. They take their personalities from the color they have, and not the other way around. Also, any day of the week could be a happy day, including Monday! My favorite day is Friday, though. You probably have noticed that I like red!"— I smiled, copying Elsa's famous malicious squint.

"Great, got it. On to months! What do you have to say about those?"

Months of the Year in Color

January—white
February—black
March—medium light blue
April—lemon yellow
May—soft red
June—bluish red
July—bright red
August—orange red
September—warm brown
October—dark brown
November—medium green
December—black

"You don't you think that the colors have something to do with heat? Look, winter is black and white, and summer is all orange and red, autumn is brownish, and the spring is bluish, going into red, then yellow. In the spring, you are missing the green."

"I really don't know why I don't have green there. It all looks like a twisted-distorted rainbow of a sort. Doesn't it?"

"Or an auditory spectrum of a sort!"

"Oh, now who is playing smart?"

"I get to say strange stuff too!"

"Since you are in the auditory mode, do you want to hear about time and space?"

"All I have been doing is listening. You should at least ask me to not flip the channel!"

"Ok, you don't have to listen! I can sign my story to you. There is also telepathy! Those don't work as well. Plus, speaking is good, because you can hear me even with your eyes closed!"

"If you knew telepathy, I could hear you even if you were not here, and with my ears closed."

"Thank actually happens to me sometimes. Sometimes I hear other people's thoughts, or I see them written on their foreheads. Well, I actually don't see or hear anything in front of me. It is all in the Phantom Reality."

"How can you see and not see at the same time?"

"Ok, it is the same as remembering somebody sing! Do you actually hear them sing? No. So, it's the same. It is in the same place where you experience your memories. I told you already!"

"So, do you ever check the validity? Because this just might be hallucinations."

"No, I am not crazy that way, and I am a zealous cartesian—I doubt myself mercilessly, and triple check everything."

"And?"

"And, lo and behold, it always, always comes out true."

"Confluence of thought, no less!"

"I don't know what that means in this context."

"I think, this has nothing to do with telepathy, but telepathy has something to do with Synesthesia."

"That is an unknown unknown, so unknown that it is hard to talk about it at all."

"Don't worry, and this shall pass, as Buddhists say."

"So, let's talk about time?"

Space and Time

"Remember my conversation about summer's plans with my mother? That was it! I experience time as an infinite colorless line, stretching into darkness and eternity on each side. Yet, time itself takes different shapes and forms, depending on what I need from it. If it is about future plans and timeframe, then I can see months in a row, with their appropriate colors. I hover over this row, and can stretch it in any way I want, zoom in and out, face forward or backward, to see days of the week and hours. If it is about my self-awareness, my past, present and future,—then I don't see any continuous line. I see a 3D space that grows in all directions at once, in 3D layers like tree bark, where nothing is ever lost. When I think about time in physics, I see the same 3D growth, but in every single point and outward. Angle is to radians—like time is to space. By the way, daytime is a downward, vertical white line, and nighttime is an upward black line. They are both pivotable, so that they can be facing me or from my right to left, or facing directly away from me, as they often do when they are part of the future. If I mentally rotate that funny-looking sinusoid line, it can makes me feel physically dizzy, as if someone is turning me upside down."

"My brain hurts, Vera."

"And my brain is itching, running around all the places it has to scratch, trying to keep up."

"Well, you know, an itch is the smallest degree of pain."

"Like a minor chord in music, is the smallest degree of cacophony."

"Cacophony is what I am starting to hear coming out of your mouth! Just joking."

"Should we stop talking about ideas then, and talk about people?"—I asked with a good dose of sarcasm. "How incredibly rude Elsa is becoming!"—I thought to myself. "What is she hinting on? Maybe that's her way of soothing herself in the midst of this mental quagmire?" I suddenly realized that Elsa was staring at me with a question mark written on her forehead. "This curvy wrinkle means dirty liver"—I kept thinking to myself. "Yes, let's talk about people!" —I said to Elsa.

"Do proceed. And you know who likes talking about people, as opposed to ideas!"—said Elsa, markedly irked by my question.

"Objection, your honor! Here, we discuss ideas about people."

"Granted."

Ideas About People

"The best idea about people I have for you is that people are colorful bubbles!"

"H-hm-hm!"—coughed and almost chocked Elsa.

"To me, a person is a collection of ideas that looks like a bubble, a colorful spherical projection into the world, with a transparent person in the center. Being inside this bubble and looking out into the world is like reading any book: you see in it only what you already know, most of the time. These bubbles are round and transparent, with all the colors on the inside —basically, like walking cells. When they meet each other, their surfaces touch. Some surfaces are glossy, some slimy, some dry and rough, some extremely sticky, some very permeable, and some smooth and invincible like a polished diamond, and some moist with life, like the skin of a potato. When they touch, each assumes a shape in reaction to the other person's texture, as they draw closer and closer. Softer surfaces give in and tougher surfaces persevere. As these human bubbles squeeze into each other's space, a chemical reaction starts taking place between them, as their surfaces permeate each other. With time, this reaction thickens, as they get attached and literally grown into each other, until separation would mean great pain, and finally, until they become inseparable. In this process, some bubbles are eaten alive and consumed by others, while others do the eating. Those who eat don't become bigger, as the other cell pops out its life like a balloon pops out its air, and then flops. When neither eats the other, there takes place an amazing electrical buzzing between them, as they merge in a rotating equilibrium, which then turns into a new single organism, and a new life is created, giving out energy into the world! Sometimes, when a bubble learns something drastically new, it can explode and stay in the shape of the explosion."

"I thought, that's when somebody goes mad!"—laughed Elsa wholeheartedly.

"Seeing something completely new is a mental explosion, Elsa, it feels like going mad. Most of us look at something new and don't even see it, because our bubble gets on the way. But then, that's the only way to see."

"Is that the only idea you have about people?"

"There is also the chemistry, not the bubble chemistry. And not the other one! People feel like chemical concoctions, with color and texture that I feel on my skin and sometimes in my mouth. I also feel their physical qualities, such as tenacity, flexibility, just like what we experience about any object in the physical world, but it is about their invisible essence— their personality. Some people just stink, and then I really don't need to analyze them any further—I know that they are no good. Some people feel acidic, some —like sweet water, some—like sandpaper, some gooey and sticky, and some like crusted yellow cheese!"

"Yuk! Who feels like crusted yellow cheese? I hope, it is not me!"

"If it were you, I wouldn't be telling you this! It is a woman in my neighborhood, though, and all I hear from her is her asking for help. She finds something on the spot! She literally invents it to make me busy and make me feel guilty if I don't oblige. Don't you know someone like that too?"

"I don't want to remember, or I am in denial. And?"

"If I am on my way to a grocery store, she will ask to buy her something. If I have new earrings, she will ask me to help her order the same exact pair. I can see her eyes searching around for something to ask, every single time, and that feels like crusty and oily cheese that she plasters all over me, and it goes into all my nooks and crannies, until I cannot breath. The crust is very hard and resilient, with cheese strings pulling on me like hot mozzarella on a pizza, and in my mind, I really strain my muscles to get it off: hook it first, then bend it away from me, then make sure that all the sticky cheese gets scraped off."

"Death by cheese suffocation! I don't think I can eat cheese now until you finish your book!"

"Imagine, everything in my head is associated with everything, and I am stuck, watching this incessant movie in my head, this soundtrack of life passing by. Like the most exciting and fascinating nightmare from which I cannot wake up!"

"Oh, it is not always about crusty cheese, though?"

"Of course, not. Sometimes, I melt into an infinite nothingness of bliss, just from what I see..."

"I see that you have trouble snapping out of meditation. Do I have any texture?"

"You actually look transparent to me, as if you are not there."

"I don't know if I should take it as a compliment or an insult. Or is this a case of when you have nothing good to say, just say nothing? Transparent?"

"Let's just say, I am glad that you are there, Elsa!"

"I don't know if I could say the same about you at all times. Sometimes I wish I never knew you. But, nothing can ever be un-known, once you know it. Could you please help me and go on with your story?"

"That's harsh. Why am I feeling a creeping deja vu of frustration here?"

"It's the cheese!"

I was about to get up and walk away from this conversation. It had gone too far! We were both becoming terrible. I didn't like the side of me Elsa was bringing out, while she showed me the side of her I never knew, the crusty yellow cheese side.

"You read my mind!"—I retorted through my teeth.

"I always do!"—said Elsa with her decidedly unpleasant smile that, at this point, started burrowing a whole through me, like a gooey leach with a metal muzzle that eats through human flesh.

"Elsa, you really don't even want to imagine what I am thinking of you right now!"

"I don't need to imagine, I already know!"

"If you continue like this—there isn't even a door here, you can just walk away!"

"Ok, I was only joking. Relax. I just wanted to see how far your bubble would stretch. It was only a little tempering exercise, to make sure you finish this book."

"No more jokes!"

One with the World

"Do you realize how sensitive I am? Most of the time, I feel the rest of the world as part of my body. Even you!"

"Something I have heard Buddhists say."

"I actually feel the surface of everything around me, as if my own bubble extends into every nook and cranny of everything that I see."

"Do you feel that about things you cannot see?"

"Only if I see them in my mind—then, they are part of me too. If they get hurt, I get hurt. If they are happy, I feel it in them, and in me."

"I think, your mirror neurons are on steroids!"

"Ha ha! This thought has tickled my brain one too many times! Did you know that a tickle is a very low dose of stress, so that it does not even feel as stressful, but we laugh, because laughter is a protective stress relief mechanism? That is why jokes are funny! You laugh because you get stressed from the surprise. And there is no comedy without surprises, you know!"

"Very funny!"

"You know why I find some people boring?"

"Not again!"

"There are two ways to looking at people: passively or actively. Passively— you sit back and wait for entertainment. Actively—you look deeper into people, to discover little troths of treasure within them. So many people sit back in nothingness, waiting for someone to entertain them, neither looking deeper into someone else, nor sharing anything interesting. And some people do something worse: they devalue what someone shares with them,"—I looked away and into the sky, full of melancholy, my eyes fixated into the blue infinity for a moment. Finally, I looked back: "Elsa, are you still with me?" Elsa was sitting with her eyes closed. She must have been asleep. I must have bored her to death.

"You are right, Vera"—Elsa answered suddenly, as if coming back from the dead.

"Oh, you heard me?!"—I smiled.

"Sometimes you can see better with your eyes closed. I am listening."

"Let me tell you more about people's bodies."

"Rated general audience please!"

"Nothing could make this idea dirty, since there is nothing for the dirt to hold onto!"

"Not in my world!"—laughed Elsa, opening her eyes for a split second.

What Dancing and Handwriting Reveal

"How one dances can tell me a lot about their personality and their brain. It is the same with how one eats or physically fights. One time, I was taught a dance by a dance teacher, and in the first few minutes,

I could feel how cruel he can be in personal relations, and how his thoughts move in these straight lines, limited and rigid, unable to grasp the big picture. He was a scary bubble!

It is the same with one's handwriting, which directly reflect the patterns in their brain, and even their word choices and writing style. Just this morning, I read someone's explanation about how the dark matter in the universe behaves near black holes. I never saw this person in real life, but I could see his slightly wrinkled aluminum foil strips of thought running out of his head, as if they were crooked rays of light sticking out of a head aura. They were inflexible and clumsy, and only its tiny wrinkles indicated that there might be some past agility in his thinking that would allow him to entertain contradicting aspects of a question simultaneously. Here, he was hopelessly stuck on one side."

"Good enough…"—nodded Elsa. "More on ideas about people?"

People as 3D Scans

"This will sound a bit medical. I see my body and other people's bodies as transparent 3D scans that I can mentally rotate in any way, and see how all the organs interact. If you actually zoom into our bodies, to the point where you start seeing spaces between our atoms, you will see that, indeed, we are transparent. We are walking clouds of energy that collect themselves into semi-closed adaptive systems. Something goes in, gets manipulated, and something goes out. Walking equations! Most of what we are (our cells!) may can be replaced, but the structure endures, so we still feel like we are ourselves. Once we were babies, made of one set of particles, and now barely any of those original particles are left in us! We are essentially new."

"I wish I never heard that!"

Hyper Mirror Neurons

"There is more. It is very easy for me to slip, and abandon myself, becoming another person, and seeing reality their way. It is quite stressful, because I am afraid that I will get stuck that way, or it will rub off on me in some way. Although, if it is a person I admire, I welcome the experience. Somehow, I pull their identity on me, like a rubber face mask, a mask thick as a whole person seen from a point of view of one single molecule within it. It's overwhelming! Suddenly, I find myself entrapped, as if buried alive inside an ancient liquid clock, that pushes and limits me, bending me by my head into a convoluted tunnel, turning me into what looks like a tree root, growing swiftly through a maze of metal pathways that make up a whole labyrinth. And this labyrinth is my head, literally, and also the other person."

"Oh, I have nothing to say to that."

"Elsa, if you are hiding an idea, then do share it! Otherwise, it is a crime against humanity,"—I laughed.

"Then, I am a criminal."

"Or maybe, you just don't have anything to say!"—I mocked her.

"I guess, you will never know! Or, maybe I just don't even know what I don't know about this."

"The unknown unknowns are the most dangerous ones!"

Inanimate Objects

"Vera, then, tell me something I know that I don't know."

"Do you not know, Elsa, that life is everywhere! Even the inanimate objects have gender and personality, and an attitude, and because of that—I can even get mad at them. I have kicked my tables and chairs for being on the way, I have caressed my car for driving smoothly, and I have gotten mad at my computer for not caring about me and being too slow. Somehow, I felt that they meant it, and I meant it back. Hasn't everyone done that at some time? Oh, not to forget that my cars have gone dead on me three times in my life, just before I had to give them up—selling them or returning a lease. In each case, no one could find the reason. 'It might be something electrical, but it is impossible to trace'—they said.

At one job that I really couldn't hate more—they replaced my computer drive three times! I would have never known, if not for a woman co-worker, who was watching over my shoulder, her breath droplets occasionally landing on my neck, her each breath sending a shower of goosebumps over my body and down to my toes. I asked why she was watching me so closely, and she told me that they desperately needed to figure out what it was exactly I was doing to the computer for it to have completely crashed and died three times. I wasn't doing anything, I was just really-really-really stressed, so the computer must have felt it and died! Three times.

Another time, I had to write an essay for the most boring class of my entire life. I hated it every key stroke of the way! Every five minutes or so, my laptop would freeze every time I touched the keyboard. I would have to lift off my hands for it to start working again. That is how I made it to the end of the essay. And how about my old clothes and things? It hurts me to throw them away, because I feel so connected to them, as if they were alive. It simply breaks my heart.

Although I am not a names person when it comes to people, I have names for everything else: body parts, my bags and clothes, my furniture, places where I have lived, household items and food, times of day, of the week and of the year, and even feelings that are otherwise indescribable. I had a purse that was called Kuka, and now any purse might be called Kuka, in her memory. My feet are brother and sister, and I will not tell you their names. If I do have names for people, they are not regular names or words, but rather sound clusters that look like the people they name.

Wines and Perfumes

What else? Perfumes and wines! Smelling both of them, I get a visual print out in my mind, like a bar code, with lower, middle and top notes. I actually feel that I smell the deep notes coming from the bottom, near my chin. The middle notes, I smell near my nose, and the top ones—near my eyes and forehead. These bar codes have textures and colors, I can clearly see how they are different, which makes comparing them relatively easy. I don't call them peachy and rosy, fruity, musty, or anything like that. That feels very remote, general and inaccurate, compared to what the bar codes show. White wines often have powdery texture, and they are empty in the middle. Red wines are often heavy on the bottom. Good wines have something to say on every level, and have a balanced personality, like a friend who has a lot to say on many topics, or a lover with whom you connect on many levels: physically, emotionally, intellectually, and spiritually. It is with perfumes just like it is with people and wine."

"Hold on! It's hard to keep up with you! Let me… Anything else on that note?"

"Nothing on that fruity wine note, but there is something about dreams. Sorry."

"Don't be sorry. I am curious!"

Dreams of Being a Stone

"I have dreamt that I was a stone, my entire body and consciousness encapsulated in stone: I can't move, despite my most desperate efforts. My muscles are literally and figuratively petrified. There is no breath I can take, no piece of my skin I can scratch—because I have no skin and no throat, even when they itch and crave air, and I can't close my eyes to stop seeing with my entire surface, all around me. It is a torture. And it gets worse. I am the stone, and the stone is being divided in halves, over and over again, and, as division grows, and so does the space that it produces, and I feel like I am uncontrollably growing in size!

Then, in another dream, I was very close and very far away, at the same time. It was like there were two of me in me, and the difference between scales of the two is what is most terrifying! The zoom-in version is at the level of atoms and deeper, and my zoomed-out version is galaxies away!"

"This is all music to my ears! And my eyes!"

"What, you see music too? I am amazed, you are liking this!"

"I didn't say anything. You hearing things!"

Music

"Whatever, Elsa! Then, music is next! I see and I can touch music. It has colors and textures, which are so clear that I can feel them with the tips of my fingers, and sometimes in my mouth, and sometimes as if they are part of my body, inside my head and inside my breath, inside and through me. There is no better way to express feelings. Music is not a language for feelings—it is the feeling itself. It is like looking at a red rose: the red you see does not represent red, but it is red itself. It is the same with music. Every objects is effulgent with music.   The walls, the building, the people, the dogs and cats, the sea and definitely the sky — they are all asound with music, and all of it mixes into a beautiful orchestral noise of colors that tell me of the emotional nature of their source. Even air has a sound that betrays its pure emotion. This is how we sometimes feel that somebody is coming, without hearing their steps.   That is how we know that somebody is right behind us, without even hearing their breathing. We hear their music, their particular unique noise that tells us all about them: their age, their intention, what they are made of, their temperature, and even what they are thinking—if they are thinking, which everything that exists—thinks. Even stones think, in a very primitive way, and you can hear it if you stop trying too hard to register it with your ear, and simply use your intuition.   With developed Synesthesia, you will definitely see it all in your mind's eye—all of those sounds.   Nothing can escape the eye of the mind, where all senses are merged into one, and knowledge is so overwhelmingly multi-faceted that is screams with certainty.

Even colors produce music! They resonate with each other just like musical chords. If the colors are similar, they clash, but at certain intervals— they ring in harmony! You have to experience it to believe it. But, you already intuitively know it, even from trying to match your clothes.

Have you ever seen light shining through a crystal prism? The way it splatters through the air and sparkles with its little colorful needles of a rainbow— that is how piano music feels. Major chords have a yellow color mixed with red. The key of C major is always bright yellow. Minor chords feel blue, and the Devil's fourth feels black. Green is for anything passing through G. Otherwise, it is a pleasantly prickly ocean of color that I feel all over my skin, in my mouth, and all over inside me."

"Have you ever seen some people spit when they hear a dirty word or see something disgusting?"

"They could be spitting from an ugly sound of an ugly idea."

"I just thought that those people have Synesthesia, if they do that. I often feel the urge to spit, when I don't like something."

"Elsa, sometimes I don't even know if you are listening to me."

"Sorry. You see, I can say sorry, too. Any other things you would like to add?"

Visual Static and the Invisible Color

"I do not only *hear* static everywhere, I also *see* it. I see air move, its molecules rolling along and amongst themselves. White color is never white but rainbow. In fact, I can never see one solid color. All colors of all physical objects are a flickering combination of all colors: very tiny, vibrating, running, flowing, and merging into each other. Even black is always like a low light rainbow, and never really pitch black. In fact, pitch black to me is unimaginable, it would mean not having vision at all."

"Have you ever wondered what colors that humans cannot see look like?"

"Of course, I have even dreamt of one! In a dream, I looked up to see the sun: it was in a color that shocked me, because I knew for sure that I had never seen it before. I remember looking at it, holding my breath, as a shower of little explosions of excitement ran from my scull, down to my neck, down my back, my legs, and all the way to the bottom of my feet. Breathless, I kept staring, trying to hold on to it with my eyes and with my mind, so that I could never-ever forget it.

The tiny dancing particles of light, of which the color was composed, shivered around, full of energy. I tried so hard in my mind to describe my visual sensation, hoping to retain at least those words, but realized that there were actually no words to describe it, or to compare it to anything we have here 'on earth.' How could I carry on the memory of it back into the waking world? I kept staring as long as I could, as if trying to take a picture of it with my eyes, my whole body focused on it, until suddenly it was swept away from me in one single swipe. I woke up and immediately tried to recollect the color.

In my mind, all I could see was a wall of grey. Just like in the dream, there were those dancing tiny lights, but this time they were all grey, with secondary microscopic reds and blues flickering in and out. The color looked empty and incomplete, as if it really was there, but it was me who was color blind, no longer able to see it. I felt such a terrible sense of loss… I feel it until now.

There are many things in this world, to which we are essentially blind. Sometimes they appear to us as that unremarkable wall of nonsensical grey, so we disregard them. We disregard what we don't understand, and end up not seeing what we don't know. However, those little signs are there, flickering everywhere around us. Just please believe me."

"I wonder, can you learn to see those things? Can you learn Synesthesia?"

"Synesthesia is not something you learn. It is more about uncovering what we already have inside us, of which we are simply unaware. You can start with the words that you would use for abstract things and see what comes to mind. You could also look at different letters and describe them as if they had personalities—without thinking much. Just your first impression. Is the letter A more kind or more angry? If neither, then try to look at it as if it were a silhouette of a person that you are seeing from very-very far away. Whom does it remind you of? Next time you see the letter A, you would think of that person and that personality, and letter A might start coming alive in your mind. There are just so-so many different ways! Essentially, it is about asking yourself questions, trying to be hyper-aware, and it will probably come to you."

"This sounds like Ideasthesia territory."

"It is that grey area between the two. Grey because we can't register its color. We really need to start thinking in clouds, Elsa! It's not all pixels and digits."

"Then tell me, what is Ideasthesia? And how is it different from what we have been talking about?"

"This is a really good question… that needs a really good answer. I will tell you what it feels like. You figure out what it is."

"I am all ears."

WHAT IS IDEASTHESIA?

"Let me see your ears, Elsa!"

Synesthesia Versus Ideasthesia

"No! What is the difference between Synesthesia and Ideasthesia?"—asked Elsa impatiently, slicing through the air with her hands, like two shiny guillotines about to chop one's head off.

"Synesthesia is when an input into one sense is mapped onto another additional sense. Like when you both hear and see music, or see and taste colors. However, Ideas-thesia, ideas + thesia = feeling ideas. It is feeling what you think. It is when thinking about abstract things, which normally don't have any equivalent in the physical reality, is experienced in one of your senses. The two phenomena might be related. Just like Synesthesia, Ideasthesia is probably present in all of us, to a different degree; but, for most of us, it is blocked from consciousness."

"Goodness!"—Elsa raised her eyebrows and looked down at her sheet of paper.

"I used to call Ideasthesia 'Thought Synesthesia', before the official term was invented by professor Danko Nikolic, around 2009. This is what Ideasthesia feels like to me: my thoughts come to me as objects of solid, gaseous and abstract nature, and I also feel their weight, speed, and chemical properties, all in a certain space. This space is part of me, and in it, I feel pressure, pain, and all that is normally tactile. It feels exactly like a phantom limb, an extension of my body, somewhere outside and near my head, up and slightly to the right. Once I mentally go inside it, it feels like I am floating in open space, completely weightless, with a pan-spherical vision—one that sees everything in all directions, at the same time. Sometimes I am on the outside of these amazing processes, and sometimes I am them, and they are me. I can always manipulate all that is in it directly and instantly, just like I can move my hands and legs,—

except, it is in some unthinkable and fantastic ways. This space is my Phantom Reality—let's make it an official term."

Phantom Limbs

"What is a Phantom limb?"

"Some people lose a part of themselves, like a limb—a leg or an arm. They lose it and then they feel that it is still there—every cell of it.

"Not as bad as losing one's mind! I wonder what that would feel like! Losing it and feeling that it is still there,"—Elsa's idea made my eyes freeze in a position of crystal wonder, wondering what she meant by that.

"I love milking my brain for insight!"—I screamed with excitement, hoping for the best meaning.

"Got milk?"—sarcastically retorted Elsa.

"You talk about it as if it were money."

"Whatever! I am still unhappy about the grey area we just talked about."

"Synesthesia is the Augmented Reality, and Ideasthesia is the Virtual Reality. And all of it takes place in my Phantom Reality."

"You and your surreptitious multiple realities!"

"You and your banausic fingers!"

"Thanks. I thought, I was doing you a favor by writing everything down!"

"Not with that acidic personality, Elsa, seriously!"

"What, are you afraid of your official Super Ego?"

"As I told you before, I am only afraid of saying nothing new. The rest is clouds. The type that explain things. The type that come alive in the best reality out there—the Phantom Reality!"

"I need to clarify some things. This Phantom Reality—where is it located?"

Where my Phantom Reality is Located

"I know, I am repeating myself. But, you asked. My Phantom Reality is located in the same place as memories and imagination, and dreams. When looking at myself from the inside of my brain, I see it located somewhere on top of my head and slightly to the right. It is like a reversed funnel that starts with the size of half of my head, and then grows wider, and outwards, indefinitely. Its shape reminds me of a flashlight's beam—that's it, exactly!

The first time I became clearly aware of it was when I was little, reading a large yellow mathematics encyclopedia. I just looked at a problem, when a solution came to me instantaneously. It felt as if some physical substances come together in my brain, matching each other's shapes, to form one whole. I was perplexed and did not trust it immediately. Then, I did the calculations step by step—only to get the same answer. This continued to happen to me later. I tried to suppress it each time, feeling as if I was losing control over myself, and fearing that this 'other brain' would take over. Later, I learned that this 'other brain'—my phantom reality was the best brain I had!

One time, a medical doctor was talking to me and looked right at my right hemisphere: my Phantom Reality projection, and I felt like she was touching it with her eyes. That was panic-inducing, and made me scream inside! AAAAA!"

"You sound like a double agent! You are here, and you are there in your Phantom Reality, too! I doubt you see the world the same way as I do? Do you even see me?"

"I surely don't see two of you, if that's what you mean!"

"I didn't even..."—Elsa started saying, as I interrupted her, because I couldn't stand hearing another word.

"It's not what you said, it's what you thought, Elsa!"

"What I really thought was, I wish I could see what you see."

"I must confess, Elsa, I do lead a double life. The physical reality around—I can never seem to grasp it to the fullest. It is like a huge ever-changing puzzle of gaps that I am desperately trying to fill in. It feels like a pointillist painting, like clouds of motley impressions passing by, details going in and out of focus, objects being in and out of zoom, clear and fading, time rushing by like roller-coaster air that pushes against my cheeks, or holding me in a single frozen breath of the present. It is all in bits and pieces, never really perfectly complete or concrete, and therefore—sometimes unbelievable. I guess, my Phantom Reality is constantly trying to swallow it in, so it could make sense of it, but can't— it is a Sisyphean task!"

"So, it is not fun?"

"My definition of fun is different. I eat theories for breakfast! That is FUN!"

"But is Phantom Reality fun?"

I bit my lip, and my eyes threw out some sparkles.

Phantom Reality Feels Like Magic!

"Oh, yes! My Phantom Reality is really a dream world where everything is possible! When thinking, I get this shower of electrical goose bumps that starts from the hair follicles on my head, runs through my neck, back, top of my arms, and all the way through my legs, down to my toes. I said it before! It happens often! Some things imagine themselves and come as a surprise. This is where all the meaning from the outside converges, and is distilled into dynamic patterns of shapes and interactions. It is like a huge, infinite beret hat I wear, tilted slightly, on the right side of my head, and once I jump inside, there are no boundaries at all!"

"Sometimes you make me feel like you are on something!"

"That's what everybody says!"—I laughed so innocently that Elsa frowned. I continued: "But I never-ever do anything of the sort. I don't even drink black tea or coffee. And forget about alcohol. No, it is all brewed locally!"—I pointed to my head with such rapturous pride that Elsa found nothing to say. My Phantom Reality is another universe, where there is no gravity or time, and where thought and body are swapped: thought has a physical shape, weight, speed and chemistry, but the body—has nothing. My Phantom Reality is where I think, it is where I really exist."

"I am, therefore I think…"—said Elsa, shaking her head, and pensively looked up into the sky.

"You know? I never-ever think in any language. In order to have a grip on the physical reality around me, I have to feed it into my Phantom Reality, where I process it and make sense of it, which happens through the chemical and physical reactions between tangible elements—which *are* my body in that reality."

"So, what is language for you?"

"It is just a mental candy to play with, and which comes in handy when I have to communicate with other humans. It is hard work, trying to translate all these spectacular 3D visions into strings of words."

"Do you think even in your sleep?"

"Dreams are pure thinking, so yes. Every moment of my life, I simply cannot stop thinking, because that would mean worse than death. It would mean the end of hope for everything that might be. And I plan on not changing, if that is what you had in mind."

"Are you a planner?"

What Planning Feels Like

"I cannot stop planning, because not planning feels like walking through a jungle forest with my eyes closed: my next step could always end up the one off the cliff. It is like driving full speed into the fog, without the brakes working. It can also be like that unbearable feeling that people get when they are seated toward the end of a bus or a train, with their back facing forward, while the scenery is flying backwards.

Time is tangible in my Phantom Reality—it is actually space. And you know what else? When things are not yet done and are pending, because I depend on circumstances or on others to get them done, I feel that I am attached to a terrible gigantic force by strings with huge hooks that pierce into my flesh and pull on me, relentlessly. Here, the expression 'pulling strings' acquires a whole new meaning! I feel completely stuck and in terrible pain, with my every minute movement I make. It is like being a camel who is pulled by a ring through its nose. Even worse, it feels exactly like participating in Tai Pusan festival, and being one of those religious devotees with hooks piercing through their flesh."

"So, I must be living in the dark! I hate planning!"—smirked Elsa.

"Oh, for me planning is the ultimate creation. It is like playing God. This vast unpredictable terrain of happiness expanding all around me. Although, plans do change like variable compound interest! Phantom Reality is an adaptive system: it adapts to the food it gets from the outside. Still, it is literally 'living the dream.'"

"What dreams may come, when we have…"— Elsa said mockingly.

"My answer to that has always been TO BE."

"I see. You would make for a funny Hamlet! But what do you actually see up there in your 'hat-like beam of light, slightly to the right'"?

Vision in the Phantom Reality

"As I said before, there is no chance of seeing two of you, because there is no stereo vision!   There is a chance of seeing an infinity of you, though!"—I laughed.

"Now, how many eyes do you have up there?"

"I never counted, but… It is probably one single eye that is everywhere—if you could imagine that. In the physical reality, there are two eyes, each providing a vision for the brain to create a 3D picture, in order to judge the distance.   There, in my Phantom Reality, there is this pan-vision—I see everything at the same time, in all directions simultaneously.   And strangely enough, it is not overwhelming, because I can always focus on one or several locations that interest me and block out everything else. It has no sense of right and left, up or down, back or forward. There is also another type of vision that I snap into sometimes: it is as if I can see only from the surface of a sphere, and outwards. Sometimes, it is only inwards. Sometimes, it is from the point of view of a line, or from two beams of light looking at each other. "

"Mind-bending!"—said Elsa, her forehead contorted, one eye closed, mouth pursed.

"Most of the time, however, it is an omnipresent vision: as if each infinitesimally tiny sphere that makes up everything that exists, sees all around it, and the dense compilation of the multitude of their visions make up the totality of what I see. It is quite intense!"

"But what specifically do you see?"

"I will tell you later."

"That's a none-thought, Vera!"

"None-questions beg a none-thoughts,"—I pouted and laughed at the same time.

"Very unexpected!"

"The problem is that, any time I start talking about what I see in my Phantom Reality, that is, when I try to translate it to words, it either all dissipates, or freezes, and is no longer valid. If I stop the words, the Phantom Reality slowly reconstructs the vision, as if through tiny capillaries, as if in lapse photography, back into its 3D shape. You know, for the Phantom Reality to rebuild itself, I sometimes have to stop breathing!"

"So, how are we going to do this?"

"Do what, the didgeridoo-doodlidoo!?"—I laughed.

"Goodness! You are the biggest nut I have ever seen!"

"I am just getting tired. And so are you."

"Yes, I usually would get up and leave at this point."

"No, you can't now, you are too deep into this. I am just expressing my fun! It may make no sense, but it serves its function."

"I am not a lobotomist, and if you can't tell me… I will have to lobotomize myself or leave!"

"Elsa, I thought, you were here to support me."

"Now I need the support. I just want to get to all the details, to the explanations of everything."

"It is not me! It's the fault of the words. They take too long, and I can't stay serious for so long!"

"Ok, ok, but you are going to tell me everything, right?"

"I am glad, you are the one who said it. Yes, I will. Well, I will tell you everything I haven't forgotten."

"Why so selective?"

"I just cannot tell you what I see now—that is quite impossible. As I have said a million times now, it is impossible to produce words and model ideas at the same time. You know how it feels? It feels painfully heavy. It feels like the vision is disintegrating because I am sucking its power and its physical ingredients away. I have to mentally squeeze myself and become smaller, so that the vision would become bigger in relation to me, as if we are two chambers filled with air, and one inflates according to how much the other one deflates."

"It reminds me of Scottish pipes!"

"Thinking, in general, feels like squeezing something out. It feels like singing: pressing on the diaphragm on the bottom and relaxing the top, where answer comes out. "

"You are full of it!"

"You mean air? Yes, when I breath in, I am!"

"Never mind, go on. We have no time."

"I am getting used to you, Elsa. I don't know if it is a Stockholm syndrome I am developing, but… you are ok."

"Gee, thanks! I guess, I am not too self-aware!"

"Believe me, it is much easier to live that way. Let me tell you something: when I do massage…"

"What does this have to do with anything?"

"I just had this idea pop into my head, so let me tell you before it is gone! So, when I massage someone, I can 'see' the bones and muscles underneath the skin, and problem areas are black and red, others are white and blue. Black is a real chronic problem, while red is a recent irritation. White is nothing special, but blue is where muscles have been working and still have the memory of work. I can see it all in 3D, and the areas that need the pressure appear in these circular ripples, like mountains seen from very high in the sky, or on a relief map. They are pliable, so that I can straighten them out with my hands. Sometimes, they are resilient, so I have to 'rub it in.' As soon as the area is relaxed, those ripples disappear."

"Oh, sounds like remote sensing in Geography, a satellite's eye view!"

"An alien-from-space-eye view! Now this—I don't know whether it is related to Ideasthesia, but looking at people's photos sends a wave of emotions, and I suddenly feel their darkest secrets, and feel how cruel they could be, or their vulnerabilities, and what they have suffered. It can be quite unsettling. If I see somebody particularly dirty on the inside, I feel like throwing up, and have to look away. Meeting people in real life is a complete overkill of information, to the point that sometimes it has an opposite effect—my brain blocks out all the information, and I become completely unaware, and vulnerable to their lies and manipulation.

It can get really tough in places crowded with people, like subways. In New York City, I constantly had to work on blocking out all these waves of information, otherwise, I would come to work already exhausted from the informational 'assaults' on subway. Funny! Being with good people makes me so happy that I see my screams of joy traverse the sky like music fountains at crescendo, filling up the whole horizon. Sometimes, it feels like it is the whole universe. Elsa, are you listening?"

"Don't interrupt me, I am daydreaming!"

"Oh, daydream this! I took this test for tetrachromacy, four or five times, and I passed it each time it at a hundred percent! They say, tetrachromats are people with four cones for color perception in their eyes—instead of the regular three. Now I know why I always have a terrible problem with makeup. It always changes color on me! Depending on lighting and location, the color of nail polish, of lipstick, of my skin looks drastically different. Indoors and outdoors, the objects around me all noticeably change color. It is not just the source of light, but the light is also reflected from everything we have around—even our clothes! This light scattering never allows any color to be consistent. Everything changes! It makes me feel like I all colors are a lie, and my mind is walking on air. Plus, one of my eyes always shows everything in slightly warmer colors than the other. I think, I told you that, right? All this keeps me constantly aware of how inconstant the reality really is."

"Wow! I can't even remember anymore what you said and what you didn't!"

"I guess, we are both getting old."

"It is not about being old! There is only so much that a human brain can handle. And, when it comes to nuttiness you are feeding me—let's say, I am proud of myself."

"I am proud of you, Elsa! I mean it, thank you."

"Never mind, I can handle it."

What Something Illogical Feels Like

"Ok, as I said before, everything in my Phantom Reality feels like it is a part of my body. There, I see and feel the ideas, just like I see and feel in my physical body: hands, face, legs, stomach. Every abstract idea has a physical and chemical equivalent in my Phantom Reality. Because everything I feel with my five senses immediately ends up there, I am extremely vulnerable to the environment. Especially, to everything illogical. Processing something illogical can feel like being pulled, squeezed, strangled and even cut, while thoughts and ideas simply do not fit together, crash and destroy each other, dissipate or explode—just like all that we have seen happen with physical objects. Sometimes it feels like being mentally raped!

"Stop scaring me! There is no way I am going to put THAT in the book!"

"You are right, strike that off. That's off the record. We sound like Adversarial Networks: I propose and you dispose! Ha ha!"—I exclaimed, and sighed with a satisfaction only known to the starving, just when they have eaten. Only because I was finally figuring out the social dynamic of this dialogue.

"By the way, we still have to connect all this to Artificial Intelligence."

"People in-the-know will see how things are connected, without our help. And I promise, I will say something about it later."

"I am not sure I am following your logic, Vera."

"Aren't we in the 'Illogic' section?"

"Ah, ha ha! Don't even try to get away with what you are thinking now!"

"And what might that be?"—I asked, full of adrenaline, as if in a boxing match of minds.

"I am not sure that I know whether you know what I am thinking about what you are thinking."

"This is worse than going off on a tangent, Elsa!"

"Just give it some thought."

"Ah, what do we know about pleasures of life? After all, few things are better than going on a tangent, to deviate from what is expected, and end up with one bombastic mental surprise at the end. What a great feeling it is to stretch the mental rubber band connection of where you started to where you want to be, to see how far you can go, before it snaps and you lose your mental radar. It feels exactly like launching a rocket into space, pause at the turning point of zero gravity, when all is at stake and all is possible, and then fall back to earth, coming home to your topic. All of it is pure pleasure!"

"I couldn't possibly put that in the book, either!"

"Well, my job is casting stones, and your job—collecting them. It is that simple."

"I think, we are doing really well in the illogical department!"

"Oh, some logic only seems illogical, when one fails to connect the dots."

"Maybe, in your Phantom Reality!"

"Let me tell you something. Two aspects determine one's intelligence: awareness and discrimination. The degree of awareness is not only an indication of intelligence, but also of being alive. Extreme awareness might mean genius, but extreme lack of awareness means death. This way, life and intelligence are equivalent."

"I must be dead!"

"Oh, who is being nauseatingly self-deprecating now?"

"Illogic wrings me like a wet rag, escalating to where I can no longer control my urge to escape. Because thought feels as part of my body, can feel like bones crushing into each other, or knives cutting into the flesh, or like something suffocating me.

I am especially sensitive to any kind of mental illness in people. I get physically affected: it feels like something is bending and liquifying me, or forcing me into a shape that I am not—like a chemical weapon, exactly! Sometimes, I hear a great cacophony emanating from them that extends far around, or is thrown in someone's direction, sort of like a sound weapon. The millions of tiny waves that grate like teeth and shred into nothing any mental life form that crosses their path. All of this is always accompanied by a feeling of terror and an impulse to run away. It does not have to be full-blown schizophrenia in someone to scare me. It could be Narcissism, or even extreme lack of self-awareness."

"How subtle there!"

"I did say extreme, Elsa! Sometimes, it has to do with my own illogic. There are times when I see that my thoughts cannot find a logical resolution. It looks and feels to me like a closed container, with arguments bouncing off of the inner walls, cluttered and unable to escape. Healthy logic has no physical boundaries, and exists on its own, in open space, because its components fit well together. That kind of logic is fluid and ever-perfecting itself, always looking for a composition of balance and symmetry, so even if it is not yet complete—it is already comforting. When that happens, it all looks like water fountains flowing in unison, like thousands of brooks trilling together."

"I am following."

"Did I tell you that I eat theories for breakfast?"

"You are getting old, my dear!"

"So, I did… Alright. I did, and I did it again. That just means that I am keen on saying it. Can you imagine? My brain literally feels hunger, when it has no idea or theory to chew on. It also literally itches when it gets curious. An itch I cannot scratch! I tell ya', it is an unsettling feeling."

"I myself have a brain itch that just wouldn't stop. Can you help me with that?"

"Your brain itches too?"—I exclaimed in amazement. "I would love to!"

"With all that happening in your brain—how do you stay sane, really?"

"I never said I was sane,"—I laughed.

"Oh, let me tell you, when I was getting ready for this meeting with you, I did some research. In his letter to Jacques Hadamard, Albert Einstein described his own thinking as physical shapes and forms interacting together to form a mixture of a muscular type, and without any words involved. Then, it would be tediously translated into words. Something like that."

"Really, where did you read that?"

"In the book called 'Mathematician's Mind', on pages 142-143 exactly."

"Now, that is encouraging! See, I have been telling you, there is something to all of this!"

"So, tell me more, I need more evidence for your sanity. Show me some muscle!"—exclaimed Elsa, again, with her peculiar smile, this time—as if she knew something that I didn't. That made me think of her in a completely new way: a sort of a Eureka moment of a newly found comfort zone of being protected by a higher authority.

What Thinking Feels Like

"You said muscle? Because that is exactly how thinking feels in my Phantom Reality! Deliberate thinking, especially creating something new, feels like a muscle strain behind my forehead bone, above my sinuses and eyes, somewhere inside. The muscles in the brain seem to work opposite to normal: you have to squeeze to release, so that the thinking has a chance to take its course and produce a solution."

"Yes, those Scottish pipes again! Did you know? That is how our eye lens works! The muscles pull it for us to focus on something closer, and relax—to see far."

"Maybe, there is some similarity. Those brain muscles contract and push in, like when you inflate a balloon with your breath: the lungs deflate on one side, and the emerging balloon on the other side is the answer. Coming up with new visions, discoveries or inventions looks like little nuclear explosions in the brain—or huge ones, if one zooms in onto them. They create their universe in the form of a gigantic mushroom network, but with lightning speed.

The Effect of Music

The brain activity looks and feels exactly like lightning, ceaselessly striking through the brain. This stream of strikes feels like infinitely complex music chords, with infinitely large number of notes, sounding at the same time. Listening to real music interferes with this stream of thought and restructures it instantaneously. It feels exactly like the cacophony of playing two different songs or symphonies at the same time! That's why I don't like listening to music to help me think. It doesn't. In fact, music dictates emotion, which, in turn, dictates thought. Major chords force relaxation, minor chords—force cause pain, which triggers an alarm in us, and that is why they feel sad. It is all about the degree of physical wave mismatching that make up music, which is experienced as pain."

"That's fascinating! That is why my brain hurts and squeaks in pain, just thinking about all this!"

A Tiny Operator of the Body

"Sometimes I feel like I am just a tiny molecule in my body. I feel like a tiny operator of a gargantuan machine, and it just overwhelms me how this huge transformer-like mass moves wherever I will it to, making leaps that are millions times bigger than me. It just feels like suddenly a whole mountain moved under my feet—except that the whole mountain is me, it is my body and I am in control."

"I need a mental break."

"I can talk about meditation."

"Never tried to meditate while listening to words."

"Think of what's behind those words, and your mind will follow."

"Let's try it."

"Sometimes it feels like the rest of the world is not real, and my consciousness just floats... Most people need to meditate to enter an altered state of mind. For me, it is a constant struggle to come out of meditation and be present here and now. The default state for my mind is daydreaming: it flies all over in space and time. I can barely hold it back or bring it back willingly. Doing something creative is pure pleasure! Having an engaging conversation with someone makes my brain create these 3D spirals of thought that are alive and function on their own, once launched. Sometimes, that conversation is with myself. Not boring at all, mind you!

Conversations, the flow of information is everywhere in nature and in all objects: everything is alive, and is constantly communicating. I avoid being overwhelmed by blocking out most of it, but my brain easily slips into meditation and reconnects with the ubiquitous informational flow. Most of the time, I find myself slightly in a dream."

"I am in a dream."

"And so, who is talking?"

"You are talking!"

Feeling the Immediate Environment

"Elsa, please either remember this or, write it down."

"I am writing it in the annals of my dreams. Do not disturb."

"Ok, then record me. I am just going to keep talking and trust that it will not get lost in the '*annals of your dreams*'"—I dared saying, since I was getting stressed about how to proceed at this point. So, I continued: "I often feel like my environment is part my body…"

"Oh, not again! Everything is part of your body!"—Elsa suddenly opened her eyes. "You are like this omnipresent ghost, you are everywhere! I can't take this anymore, seriously!"

"It is what it is. Let me just tell you, and you decide what to do with it. When I look at objects, I feel myself wrapping around them, so that I can know and feel their 3D dimensions. Sometimes I see myself as an object, and observe the environment from its point of view. What I see often shocks me and even scares me, to the point that I have to force-snap myself out of it, as if from a nightmare. If an object is broken or cut, I feel the pain, or various types of discomfort. It all often gets overwhelming, and my brain just zones out.

Moreover, the objects I observe in physical reality can feel abusive because of how intimate my experience of them is. When I see something dirty, I can feel it on my tongue, or sometimes smell it as if it is right there in my nose—to my great disgust!

I also easily get disgusted with ignorance, especially when it is combined with pushiness. The same with injustice and immorality, because they become this material that touches me everywhere. It gives me a bad taste in my mouth, and my skin feels terribly gross all over, and I try to un-see and un-hear things. It is pure panic."

"Oh, I feel so sorry for you, ha ha!"

"Hey, there is more. You will like this one."

"Why are you so sure?"

"Because I like it. It is fun! So... I like to talk gibberish sometimes."

"Oh-ho-ho! I never noticed!"

"Ah! It's not what you think! It only sounds like gibberish, but in fact, it's a texture language, and the sounds I make reflect the textures and colors I experience, creating words and mixing grammar from different languages. Here is one quick example. One of my favorite words is 'Siasia!', when someone comes home, and I have been home alone for a while. And this has nothing to do with the verb to see, as is 'see ya.' It is all about the burst of excitement that I literally see coming out of my head that starts with a sound 's' for thin, tiny and reddish in color, grows into the sound 'ee' that is blue and intense, and explodes with a 'ya' part that is bright yellow and white, and streaky, coming out into all directions. "

My Phantom Tail

"So much information. Anything else?"

"Unmistakably, politeness is not one of your talents, Elsa! Well, I can tell you about how I miss my Phantom tail!" Elsa's eyes, apparently, were connected to her brain with chewing gum, because I could see it hanging from her face, five inches out, to where her eyes were right at this moment.

"Ha ha, I love your reaction! It means you care!"

"Now, that's... new! Go on."

"So, I miss my phantom tail. I would love to have a prehensile tail, so that I could hug someone and stroke their hair at the same time, so that I could look for the keys in my purse when both of my hands are holding groceries, or could insert an extra note or two when my hands are busy playing piano. I could also answer my mom's calls, when I am baking bread, and my hands are covered with dough. I could also attach a camera to the end of my tail and peek into places that are otherwise out of my reach, like the inside of a huge old clock."

"You can't miss it if you never had it!"

"Well, I feel that I was meant to be born with it, or maybe, I once had it, then, somebody chopped it off, and now I am left with just a memory of it and an unending nostalgia."

"I sincerely wish you to get your Phantom Tail back, Vera! I think, this will be all for today."

"Did you get mad at me, Elsa?"

"I am not mad, I am feeling something else, I don't know how to describe it…"

"It's ok, I don't really want to know."

"Ignorance is all about fear."

"Elsa, once again, it is not what I said, it is what I thought that matters."

"Same old violin? Ignorance is how good intentions convert to evil actions."

"Is this a competition in insanity?"

"It is nothing more than a good old revengement."

"Revengement? At least, you are not boring."

"The glass is only half full because of gravity."

"Ok, I will think about it! Let my Phantom Reality figure it out."

"You know, Vera? Despite everything, I enjoy listening to your nonsense. I see a pattern in it, a pattern of insanity that promises to make sense."

"I only need one single person to understand me!"

"I am starting to understand…"—smiled Elsa. "I just need to make sure that my understanding fits into a standard book that you will write someday."

"Ok, I will try my hardest. This was all just an emotional brainwashment to pave the way for the main brainfly."

"I think, we speak the same language. I think, I will just flap my ears now and fly away!"

"Wait until the next section!"

CONFLUENCE OF THE SENSES

"Elsa, now that we are in the future section, you did tell me once that you were all ears! Wouldn't it be reasonable to imagine that you actually have no brain?"—I said coquettishly, looking straight at Elsa's tired eyes. I was starting to like this game.

"There is no reason in imagination. You could also imagine that I already flapped my ears and flew away. And I actually could do that!"

"That's right, but I need you, Elsa."

"That is precisely why I am here, despite having flown away in my imagination."

"Do you see how fun it is to have a brain?"

"I thought, I didn't have one?"

"I guess, you do, if you can fly!"

"We might speak the same language, but we probably do not think alike."

"The difference is superficial, trust me. I think, Ideasthesia is how everyone's brain functions, but for most of us, it is pushed so deep into the subconscious, that we are simply unaware of it. High awareness of Ideasthesia might have been the early human condition, but got de-selected by the evolution, because it can easily cause insanity, if the brain is not 'muscled' enough to handle the experience. In a place where the selection is in favor of those who can run the fastest or hit the hardest, the hyperawareness of Ideasthesia is a disadvantage."

"Yet, in our times, we can afford entertaining it, because there are no lions lurking outside our front door."

The Five Senses as a Single Channel

"There are other kinds of lions, lurking in other places, in our otherly times. But, let me tell you more. What I see in my Phantom Reality is that all five senses used to be a part of one single channel, in those primordial times when homo sapiens did not yet exist.

91

It appears to me that later, the senses divided, because organisms with multiple ways of complementing information about the space around them were more evolutionarily fit. The senses seem to have divided differently in different species, and there are more than just the five senses that humans have! Finally, it might be an accident that humans have those specific five.

Bees are sensitive to the electromagnetic field, for example. How about echolocation in dolphins and bats? I wonder, how they experience sound: do they actually see it? Snakes actually see with their tongues! That is why their tongues are split: for the stereo-vision through taste! I wonder how creatures who navigate physical space through hearing, taste, smell and touch actually experience reality in their brain. Their experience might not be drastically different from vision, except maybe, for the absence of color? Or it may even be that color is only a construct, a sort of a hallucination in the brain and is not exclusive to vision, but can be triggered by any sense. It may be that color is only one of the brain's tools for modeling reality, which is then experienced as consciousness.

With us, humans—if you close your eyes—you visualize the space around you, and can actually orient yourself in complete darkness, even though it might not feel the same as primary vision. When you speak and hear your own voice bounce back from the objects in your environment, the echo paints a picture of the space around you. Blind people are known for using their voice to move around."

"It's true! When I come close to the mirror and close my eyes, then speak to my own reflection, I can hear myself right in front of me."

"If you think back into the very beginning of life on earth,—our senses are all about our movement through physical space, which allows us to eat and procreate—the two main functions to ensure survival. Our senses also allow us to avoid dangers like predators, pernicious chemicals, sharp objects, and poisonous foods. Our sense of vision uses light to model our environment in the brain, and is our most powerful sense to help us do that, but could the other four be used for the same purpose? It seems that our senses work in synergy, like checks and balances, triangulating each other, correcting and complementing each other's input to help us read our environment more accurately, and therefore be better fit for survival."

"That might explain why in Synesthesia, the senses are intermixed—an atavistic feature, reminding of our simple sensory past. Also, when dolphins and bats use echolocation, which is sound, to see—isn't that also a kind of Synesthesia?"

All Five Senses Work with Waves

"Exactly! Also, if you think about it, all of our five senses work with waves! Vision is enabled when rods and cones process light waves, hearing is attuned to the sound waves picked up and conducted by the ear drum, touch picks up the patterns of movement that are made of waves on the atomic scale, while taste and smell read the chemical signature of substances comprised of waves. It all points to the possibility that all five of our senses come from the same single neural channel, originally. Plus, I just see it in my Phantom Reality. So, I am sure of it."

"Our tactile sense tells us a visual story of what is around us too!"

"Exactly! To think of it, how do you experience touch? To me, it is primarily visual. Try it: close your eyes, and touch different textures. Do you see them, or do you start hearing them? I would imagine that most people without Synesthesia will see and not hear. Smell and taste, likewise, seem to cause visual experiences, albeit very blurry ones. If we close our eyes and smell someone's strong perfume, as they pass by—we clearly smell that person as a moving object, and in addition—we also smell the time borders of when they arrive and when they leave. It is still visual! Taste is blurrier, but tasting something tells us that it is indeed there. Put some plastic in your mouth, or something that has no taste at all—and you will see how discombobulated and unsure your brain becomes about the object's existence!"

"I think, all senses are fundamentally tied to vision, since it is vision that serves us best at navigating physical space. That way we stay alive long enough to procreate and pass on our genes."

"I think, the five human senses are quite limited, as compared to what they could be, to how they function in other species. The senses we have are just enough to survive, being humans. We don't directly perceive magnetic and radio waves, for example. Our vision itself is within a certain spectrum of waves, and we are absolutely blind to the waves that fall out of the visible region.   We have a hard time believing things that we don't sense.   There is so much anecdotal evidence for telepathy, for example.   The sixth or the seventh sense—could it be there in some of us?"

"We know so little that we barely know what we yet don't know.   And even when we know what we don't know, we would rather call it impossible than admit that we don't know it."

"It was Socrates who said 'I know that I know nothing.'"

"I don't even know that much!"

"There is also an ancient Arab saying that goes something like this:

> *'There are people who don't know that they don't know,*
> *There are people who know that they don't know,*
> *There are people who don't know that they know,*
> *And there are people who know that they know.'*

"So, which of these are you?"

"Well, I would wish to know the unknown of which I don't even know yet."

"Ah, the unknown unknowns—that which lies outside of all our senses, and even our imagination."

"Maybe I don't really want to know, or I would immediately get bored with it, once it becomes known."

"'An unexamined life is not worth living'— Socrates knew that."

"It is true. After all, we don't even know how many unknown unknowns there are out there."

"All I know is that it never ends…"

"Elsa, you are so different now!"

"That's because you don't really know me."

THE BATTLE OF THE BRAINS

"Elsa, I am sad. I am sad, just to think about how much I still don't know and will never know."

"On the other hand, think of all the things you already know that you still don't know. Just focus on that! The unknown unknowns will make themselves known, with time. Also, against all odds, it is your imagination, the extended antenna with a little eye at the end of it—your phantom tail— that can reach to them!"

"Elsa, my brains are having a fight!"

"Oh?"

"Please put that in your notes. This is important. In the fight of the brains, it is not the one on the top that wins!"

"Got it? But isn't there a lot more that you are not telling me?"

The Inner and the Outer Brain

"Yes, there are at least two brains in my head, like a partitioned hard drive. One is on the outside, and one on the inside. The one on the inside is also somehow lower. So, the top is the same as outer, and the bottom is the same as inner. They roughly correspond to the conscious and the subconscious."

"I certainly didn't know that! Is that even spatially possible?"

"In my Phantom Reality, contradictory things are possible, because I can see all alternatives at the same time. So, the top-outer brain is a control freak, it constantly tries to stabilize thinking by creating blueprints, by tightening regressions, narrowing down the scope, limiting thought processes, so it is easier for it to operate. It constantly observes and controls the inner brain—at least, that is the illusion it entertains for itself. The bottom-inner brain does all the work and is the one who really knows everything. If we ever feel like we have conflicting thoughts or feelings—it might be that our two brains are fighting.

The worst situation is when there is a misalignment of how much each one understands. When the outer conscious brain understands more than the inner—it feels powerful and on top of its thinking ability—which is, at least partially, an illusion. The inner brain understands more than the outer—it is that gnawing feeling of knowing something, but not being able to explain it. Like a word on the tip of your tongue. In this case, the outer brain feels confused and incapable, although it is only a sign that it knows more than its awareness allows it to realize.

Between the two brains, there are multiple opinions. Some of them arise from a synergy of two or more impulses, and the synergy of synergies, resulting in a vast hierarchy of thought that rises lightning-fast, in response to a simple stimulus. I see these millions of vectors 'regressing' and uniting into a smaller clusters, growing into an intricate hierarchical tree-like sponge that keeps historical paths for each vector. This is what emotions and thoughts are made of. No wonder, we are often confused!"

"'If you are not confused, you are not thinking'—used to say my philosophy professor."

"Or feeling!"—I laughed. "These two brain really remind me of the Adversarial Networks in Artificial Intelligence. They are like us! It's also like writing and reading modes. In writing, you diverge and create more or less indiscriminately, not noticing typos and logical errors, because your focus is on the production. The outer brain only provides clues and suggestions, while the inner brain does all the creative work. Then, when you read your writing, you focus on receiving information, and you cull out the unnecessary. This is where the outer brain shines!

And this is why you usually don't notice your typos until you read your writing later: at first, the outer brain needs to relax and let it go, so the inner brain can be free to create the writing. When the inner brain is done, the outer brain takes over to monitor the mistake detection. One always dominates. It is also the reason we so often do much better when we don't care: it is because such an emotional state releases the grip of the outer brain over the inner. The sanctioning machine of the outer brain is silenced, so the inner brain can do its work freely. It is so not only with creativity, but also with solving the most difficult problems that deal with the unknown unknowns, and even with learning foreign languages. This is probably why creating something often feels like squeezing a balloon on one side, and releasing the other side. It is when the inner brain works and inner brain relaxes.

In addition, the two brains do their thinking differently. The inner brain sees the whole picture and all the factors at once, including emotion,—whose value is attached to every entity, event, and calculation it does. It prepares and packages information for the outer brain, from which the outer brain inductively gleans its logic, so that it can think, too.

Although logic is considered objective, human logic is only based on the limited information provided by the inner brain, all of which is inherently rooted in emotion. That also may be why we can understand emotion intellectually, but we could never feel our way through logic: because logic is always secondary in our thinking. Just an idea!

This also reminds me of how it is always a strong emotion in me that gives rise to the best of logical thought, and a lack of emotion—stifles all thinking. How curious!

It's quite pitiful how the outer, less powerful, brain is pushing the inner brain around, especially in creative tasks. When we say 'it's art, it is not science'— it only means that the inner brain has figured it out, but the outer brain still cannot understand it.

We associate our true self with the outer brain, while the inner brain, which is the source of our intuition—well, we treat it with condescension and denial, even when it is much faster and efficient than what we consider our true self to be. We think our intuition is not the real 'us', and is inferior; yet, it is our inner brain that knows more than we could ever know, given the brain structure we have.

For example, how often do we trust our first idea, when answering a test question? Statistics show, and teachers tell us, that it is the very first idea that comes to mind that usually is the correct answer. Yet, we don't trust it—because it did not come from our true self—the outer brain, and we don't remember doing any work for it."

"Do they feel different?"

"Yes!"

Inner Brain and Intuition

Top-outer brain is mostly what we associate with our directly analytical ability, and the bottom-inner brain—with intuition. The intuitive system of thinking is quicker in its decision-making than what we call analytical, but both systems are indeed analytical. They really both do the same work—they analyze information, only the intuitive system has a gargantuan scope and its workings are mostly unavailable to our consciousness. The intuitive system feels more accurate, but tests to be less accurate, I have read, than the conscious analytical system. If it tests less accurate, it is certainly because it is under-informed, under-trained or under-used. Its lower accuracy is not inherent in its processing, but in the limited or inaccurate information it is provided, and it is systematically suppressed as inferior and not exercised, unfortunately.

Maybe it is because we humans are barely starting to understand it, and in our ignorance, categorically discount it. All the while, the intuitive system is googles more powerful that the familiar conscious system we use, and would give us incredibly more accurate and telling results if we would give it credit, and then develop it and inform it better. The top-outer brain system is just a flat, small scale, zero-in, focus-on, super-simplified version of the intuitive system for our limited conscious field of focus, while the intuitive system mostly works in our subconscious so as not to overwhelm us. Imagine seeing in your mind's eye all the data-mining, all of the associations, all of the modeling and testing, checking for patters, shuffling and merging and alloying—all of which happens at the same time and in different directions! We would go insane! This is what Ideasthesia feels like at times."

"Is thinking equally clear for both of the brains?"

"The outer-top one feels like a foveal region in the retina—a small patch of clarity in the midst of a sea of blur. Like a tunnel vision: very limited in what it can focus on and see, simplistic, and lazy. The bottom-inner brain feels like a hard worker, heavy, deep, like a huge boat, a Titanic breaking through ice and carrying a heavy load, reaching down all the way to the bottom of the ocean, processing and seeing more than one can see—all at the same time.

Multitasking

The top brain cannot parallel process anything—basically, it cannot do two tasks at the same time. Multitasking is an illusion! All it is—is switching very quickly from one thing to another. Do you know, when we look at someone's face, we cannot even focus on both of their eyes at the same time?   The true multitasking is done by the bottom-inner brain—it sees and considered everything at the same time, while any smallest part of the process is aware of what is happening in all the other parts."

"Can we train our top-outer brain to multi-task like the bottom-inner brain does?"

"I don't think so, because it goes against what it is built to do. It is meant to focus on one little thing at a time. Teaching it to process more than one thing in parallel is like making an object be at two locations at the same time. Only the bottom—inner brain can do it, because it has the structure for it. In fact, forced multitasking deteriorates the quality of our thinking by inducing a fake state of alarm, which makes the fight-or-flight response to take place. It causes us to suppress the bottom-inner brain's activity, and rely on the top-outer brain. We start skimming the surface, using shortcuts, stereotypes, and heuristics to interpret new information and make decisions. In this state, people are at their most vulnerable and easiest to manipulate, as they react predictably to simple clues, without being able to evaluate them within a big picture. I try to stay away from multitasking and really rely more on my bottom-inner brain a lot more. I trust it more, I cannot even think without it!"

"No one can! I understand how some people try to claim in court that they are not guilty of a crime they have committed, blaming it on their brain!"

Responsibility for a Crime Committed

"You mean, they say it was not them, because it was their bottom-inner brain?"

"Exactly! Since they are not fully aware of what the bottom brain is doing, and they can't fully control it—why would they be responsible for their actions?"

"Because their outer brain can sensor and go against what the bottom brain is telling them to do. So, for example, if you like a certain candy, and you

107

like eating it without paying for it, your top-outer brain still knows that it is wrong. If you feel your desire stronger than other people—you are just out of luck with nature. If one is aware of what is right and wrong—one should still answer for their crimes, because their top-outer brain knows better. In the end, the outer brain is one's free will. And remember, humans consider the top-outer brain their true self."

Phantom Reality and the Two Brains

"How does this all relate to your Phantom Reality?"

"I think that the Phantom Reality is how the bottom-inner brain communicates with the top-outer brain. The inner brain just does not have any other way of expressing itself. Some of us are aware of it, and so we experience all the shapes and textures of our abstract thinking. Some don't see anything at all. But it is there in all of us—that is what I think."

"What part of your brain thinks so?"

"The only part that does all the work—the bottom-inner brain. The top-outer brain only agrees with it, since it doesn't really know."

"So, which one of them is your true self?"

"For me, they both are. Just like each of us here is a true friend of the other."

Elsa smiled with the warmest and biggest smile that I had seen in our whole conversation!

"Between the two of us, who is the lazy one?"—asked Elsa.

"This actually does not match at all! I am the lazy one, but I am the one who is doing all the thinking and talking. That makes me the inner brain,"—I said as if waking up from a daydream.

"But I am doing all the writing—does that make me the outer brain?"

"Well, compiling and writing down what somebody thinks and says—that is typically a top-outer brain activity."

"And people think that it is the top-outer brain that is their true self, right?"

"That means, you are the true self of this conversation!"

"Let's leave it to be an unknown unknowns."

"How about mushrooms?"

"Now—what?"

HUMAN BRAIN IS A MUSHROOM

"Elsa, you know well that the brain resides in the ears, right? It also resides in the eyes, and in the skin, and all over the body. The body is just an extension of the brain—its sarcophagus."

"I think, you mean to say—its exoskeleton."

"That's right. I see people's bodies as gigantic Transformers, each with a little tiny mushroom, sitting inside their skulls, ruling from above, using the body as an enormous shell to wield its will on the world and to protect itself."

"Brain is a mushroom. Hm, curious!"

"There is actually more than one, it's a family! The first mushroom is the inner brain, and the second mushroom is connected to it via something that looks like an umbilical cord! It is the top-outer brain that is watching the first one. The third mushroom is probably where we experience meta-cognition, and is the part that is watching the consciousness. The next one is watching the meta-cognition. They are all connected. This reminds me of calculus: location, speed, change of speed, how fast the rate of speed changes, and so on ad infinitum. What will happen if we keep adding more and more mushrooms?"

"It would mean omniscience. I feel like a mushroom myself now!"

"You are a mushroom! Because you are your brain, right?"

"I meant it in a bad way"—said Elsa with a frown.

"Ok. What I see in my Phantom Reality is that a human brain as a whole is a big extended family, with distributed cognition. If you look closer, and zoom into what the brain really is—you will see a whole mushroom society, all seated inside their Transformer shell. Have you heard of mycorrhizal networks?"

"Yes, that's where mushroom networks connect plant roots underground."

"That's what I am talking about! To create a brain—we have to grow it, just like we grow a mushroom, or a mycorrhizal network, and just like we grow a human population."

"Of what material?"

"I know that this will shock you, but I think, it could be anything, even a solid crystal, or even water. It just has to be grown and not assembled. That would be like putting pieces of human body together, hoping that when you merge them, a live person will emerge."

"Have you noticed that a lot of mushrooms have exactly the same color as the brain matter?"— Elsa whispered in horror.

"Ah, the millions shades of grey!"

"If it is a mushroom, how did it get inside of a human body?"

"This will sound like complete heresy, but! There is a possibility that somehow, the brain is a result of something that functionally appears to be an infection. Maybe a fungal infection? A mushroom infection? Structurally, the human brain acts as a 'farmer' of the rest of the body, in that it uses it strictly for its own purposes. Think of the blood-brain barrier:

it is as if either the brain or the body—one of them is protecting itself from the other.

It may be that the embryo is also seen as an intruder to the female body, and therefore women get nausea and other symptoms of intoxication, until the body purges the embryo through birth. What if the brain is the main reason that the body reacts this way? Maybe the brain is a mushroom that is recognized by the mother as something to be rid of, before it takes over?"

"Your imagination knows no limits."

"You are a good listener, so I simply obliged. Who else would listen to my zany ideas? That is why we are a perfect Adversarial Networks pair, Elsa."

"And all I do is dispose, like a garbage disposal for your insane ideas."

"Sorry."

CONSCIOUSNESS

"Vera, before we proceed, we should calculate your MLT ratio."

"What is that?"

"MLT stands for meaning per line of text. It is a ratio, obviously."

"Anyone speaks Phantom here?"

"Didn't we already say that we speak the same language?"

"Well, if we did, you would understand me. You would understand that we do need these wild leaps of imagination for scientific purposes, and you wouldn't be trying to establish my MLT value."

"Since I dispose, as an Adversarial Network, most of what you say, I will dispose, before I include anything in my notes."

"Go head. Just remember—this book will have been written under duress. The only reason I have you take notes is because if it were for me, I would keep it all."

"I feel like that theory of yours, about a mushroom being an intruder, making a pregnant woman nauseated—that is working on me now. You make me feel like I have one of those mushrooms inside."

"Ok, ok, if I tell you about what I think consciousness is, maybe, just maybe, you will rate my MLT higher?"

"Just brainstorm here, I will do all the automatic cleaning. So, what is consciousness?"

Consciousness Continuum

"This what I see, in my Phantom Reality. Consciousness is a degree of awareness of self and the environment, experienced in the outer brain, and Ideasthesia increases this awareness by including the inner brain into the picture,"—I started nervously. "It is in the movement of information and interaction between the inner and outer brain that we experience our being alive.

The degree of consciousness is distributed along a continuum: from elementary particles, to stones, to plants, to animals, to humans, all the way to a superconsciousness that returns to elementary particles, closing the circle.

In humans, it varies among individuals, as well as for one individual. Some people have a very poor self-awareness, and are unable to examine their own mental states and behavior. Others are hyperaware, making their Emotional IQ high. Human consciousness increases during times of danger, when everything seems to happen in a slow motion. It decreases during times of high engagement in an activity, drastically decreasing during sleep, and going down to almost nothing during anesthesia—during all of which the inner brain is deactivated and the inner brain is active."

"Wait, does that mean that we are less conscious when we are happy? But, are we really conscious when we are asleep or under anesthesia?"

"It depends on how active our outer brain remains. Our inner brain never sleeps, but it is not where we experience consciousness. If there is a loud sound in the middle of our sleep, we hear it with our inner brain first, it wakes up the outer brain—and then we are awake."

"And under anesthesia?"

"I think that we are aware of a lot more during the procedure than we can remember after it is all over. In a way, that points to a special kind of a consciousness provided by the inner brain."

"You know what makes the outer brain a special contributor to our consciousness? It is our outer brain that allows us to feel rich or poor, plan our next vacation, try to win in a chess game, and sacrifice our life to prove a point. These are all abstract constructs that we mentally superimpose over the physical reality. Although the inner brain helps with processing, it just wouldn't care about such things, unless the outer brain gave it the direction. Yet, for the outer brain, it is all about seeing a framework, buying into it, and then playing by its rules—a non-existent virtual pattern detected and accepted as a challenge. It is our outer brain and its meta-analytical function that makes up the chasm between us and the animals."

"I am still puzzled about how an individual self-awareness can change, outside of sleeping or anesthesia."

"Syneshthesia, Ideasthesia, and... anesthesia. Anesthesia literally means absence of senses. Interesting."

"Please, answer my question."

"Yes, this could have something to do with how fast or slow the passage of time is perceived to be.

Perceived Speed of Time

The speed of time in our perception is conditioned by how alert the outer brain is. The more alert the outer brain, the slower the time seems to pass by, as when you are fighting off a danger, running away from a predator, or experiencing a car crash. It is as if the outer brain starts recording more frames per second, stretching out the time, to become more attentive. On the other hand, during daydreaming and safety, the time speeds up, as the outer brain relaxes and passes the reigns to the inner brain. Deep sleep is where the time passage is the fastest: you barely feel how those eight hours pass by, if not for selected dreams. When you are having fun—the time goes by quickly, when you are bored—it just cannot go fast enough. That might have to do with the outer brain and its deliberate analytical function being disengaged during times of pleasure, and what is often called 'the flow', a term coined by Mihaly Csikszentmihalyi. The time seems to go slower when we are bored and hyper attentive, being in an active state of searching for 'the flow.'"

"That means, time is relatively relative."

"The brain's theory or relativity!"

"And all of this you saw in your Phantom Reality, right?"

"Like that reality—there is no other. It is its magic realism that tells me all I ever wish to know. And things I never knew I didn't know."

"Why do your eyes keep jiggling, as if they were looking inside your brain?"

"They ARE looking inside my brain, how did you know?"

"Because they don't seem to focus on anything that I locate in space. Yet, it is clear that they are looking at something in detail. I really want to understand it."

"If I told you, I would have to make you disappear!"—I laughed.

"Just tell me and see if I disappear on my own!"

119

THE EYE MOVEMENT

"Elsa... you still here? Do you remember the first time you went to see the ocean?" Elsa looked up and to the left, rolled her eyes around a little, then looked up and to the right: "I can't remember..."

"And I remember everything. It was the northern Baltic Sea, as cold as a thousand strings, viciously slicing your skin, or like piranha teeth tickling you, before they eat you all the way to the bone. I also remember how, when my legs got numb from the cold, and the sea pulled me in, not to drown, but to tell me something. I realized that all the water in the ocean was calling for the water that was in me, to come back home. Humans came from the sea, you know, and we all consist mostly of water."

"Why were you looking up and to the right, when you were telling me that?"

"My eyes were flying to that distant time and far-away place. And why were you, too, looking to the left and to the right? I think, your eyes were searching for something, for that memory that you did not find and then imagined, but were too shy to tell me about."

"It really did feel like my eyes were flying away!"

"See, your eyes can fly! And eyes don't lie."

"I am my eyes, after all."

"I know where you were flying—your own Phantom Reality! Have you noticed how, when people daydream, remember or imagine something, how their eyes start moving, as if they are groping some crevices of an unknown terrain, focused on something that clearly does not exist in space in front of them? They are searching within their Phantom Reality, where dreams, memories, and all imagination lives. You can tell a lot about people's character and intelligence by watching their eyes when they think."

"What can you tell?"

"The more focus and energy there is in their eye movements, the more intelligence there is in their brain. It also seems that blinking rate is revealing of how fast the people are switching around from one aspect of a big picture to another—which in itself is not necessarily indicative of high intelligence. When people listen to music or watch a film, it seems that they blink with their rhythm. Everything has its timing and must work just like a pendulum in a clock. Human heart is a type of a pendulum, pumping blood around the body: up and down."

"Maybe a yoyo? People even gesticulate to follow their eye jiggles when they are thinking. There is nothing there in front of them to be searching, yet the eye move as if they are searching a long a complex terrain. This makes me think of the rapid eye movement in REM sleep. Maybe this is like REM, but with one's eyes open?"

"Maybe thinking is similar to REM, and REM is really thinking?"

"You know what else? When people sleep, they are all cross-eyed, under their eye lids. Yet, when they think intensely, their eyes are focused on something infinitely far."

"Or, on something inexistent in front of them."

"Right!"

"Left! It is whatever direction. "

"I wonder what those directions are about. We probably need a 3D map of the Phantom Reality— that will tell us what's where!"

"Many times my Phantom Reality looks like the inside of a cave. You can close your eyes and walk through it just with the sound of your voice, reflected from its surface."

"Just like bats!"

"By the way, bat voice sounds like a mix between a mouse and a bird! I wonder, could we ever understand what they are squeaking? We can probably start with their intonation. By the way, people's intonation can tell a lot about who they are. I see intonation as a print out of textures and colors, objects that have chemical and odor qualities."

"Does that happen to you with bats too?"

"In terms of experience—yes, but I haven't heard enough bats to know their character! Although, I am sure, from the little that I have heard, that it works the same with bats as it is with people. And dolphins. Or anything that communicates with sound."

INTONATION

"Intonation is just like one's handwriting—all your emotions and intentions are there, including you level of intelligence, and…"

"And every thought you have?"—laughed Elsa.

"And much more, really."

"Tell me about it!"

"I will, if you stop laughing at me, and tell me that somehow you will include this in our book."

"Oh, it is *ours* now?"

"Oh, just put your name on it, it is yours, since you are taking all the notes."

"Intriguing, albeit a bit trite, in the context of our conversation. Never!"

"With that kind of an intonation—I know who you are!"

"I have been listening to you more, though. Hence, I know you better than you know me! Heh-heh!"—Elsa pointed out, with one of her eyebrows flipping up so high, I thought it was going to cross her hairline. I quickly realized that her determination was winning over mine, and we had to proceed onto another topic, before I lost all my positions. So, I smiled with a smile of a girl right out of the finishing school and uttered ever-so-pleasantly:

"How is the weather, dear?"

"It is just fine. Couldn't be finer. Nothing more about intonation, I see?"

Elsa was writing all this time on white printer paper. She would not type on a computer, she said, because it would not engage her fine motor skills, which she said trained her intelligence. I looked at Elsa's hair, while she was looking down: it looked so much like mine, just longer. I thought of how much I would love to grow my hair to that length. Maybe, just maybe, if I would pull on it—it would grow faster? It is like if I talk faster, maybe what I say will become more meaningful. Elsa kept looking through what she had already written. I waited a few minutes and then asked her:

"Did you notice that I am still sitting in front of you?"

"Uhm…"

"The 'uhm' is like yawning: it very contagious, did you know that? And I really don't want to catch that wordly disease."

"Is it, really? God forbid! I must have caught it from someone. Is there an wordly antibiotic for it?"

"Yes, listening to someone who does not say it and staying quiet, so you don't make them catch it, then catch it back!"

"Oh?"

"Really, that's all it is. But the 'uhm' disease can quickly become an epidemic, when the speaker has a large audience. I have seen this many times: rotating speakers, many people watching. One speaker comes up and says 'uhm' a lot—the speakers afterwards will do the same. It happened to me, too. Appalling! I am in such a terrible pain from it! It is a torture bad enough to make me confess all my credit card numbers. That pitiful thing makes me get into the brain of that person, and feel that I am them, and then assume their mental struggles—because that is what 'uhm' is a sign of!"

"Just calm down, please."

"Ok, ok."

Intonation and the Upbringing

"And?"

"Intonation can tell you about someone's social status, even in a person who speaks a foreign language you don't know. What is amazing is that poor education sounds the same in all of the world's languages: it feels sticky, murky, gooey, with irregular borders, grayish with blurts of colors mixed in, and just plain ugly. It tastes bad in my mouth. It feels like it forces my whole body into a certain shape, dragging me on the ground, into something I don't want to be in.

By education I don't mean a certificate of graduation, but upbringing. Not even that! I mean— the good nature, the integrity, the kindness of a person —that is the true sign of a high social status, regardless of how much money or power they possess, or how much all of their graduation certificates weigh, combined. Bad human nature sounds aggressive, indistinct, uneven. Good nature smells fresh, feels smooth and cool to the touch, looks symmetrical and transparent. Those qualities sound the same across all

127

languages—I never heard an exception, in close to a hundred languages I observed!"

"I think that intonation also varies by language."

"This is completely politically incorrect, but I will say it. You know what I see in my Phantom Reality? Intonation, inherent in a specific language, influences the freedom of thought in the population who speak it."

"Something like Sapir-Whorf hypothesis?"

"Maybe. Yet, it is not about the word choice, but the degrees of freedom that the intonation allows: how much one can vary pitch, the melody of the phrase, the rhythm, without sounding strange. Some languages have a very repetitive, rather flat in pitch, and inflexible intonation, and I think that somehow that curbs how free native speakers of this language feel to be different, and to create something new."

"Risky. But interesting!"

"A litmus test for it would be learning it as a foreign language, as an adult, and watch how it makes you feel. I have specific languages in mind, but I won't mention them."

"Would that affect one's intelligence?"

"Not really. Only in so much as it would either encourage or discourage venturing into new territories of thought. It is the invention of something new or deviating from status quo that seems to be affected."

"By the intonation?"

"Yes, by the intonation."

"I will have to listen to mine now!"

"I am worried of mine too, to be honest... Oh, and the politicians! Have you noticed the change in the last hundred years or so?"

Politicians' Intonation

"I haven't been alive for *that* long!"

"Right, I am only five, but I remember. So, the politicians used to have a very different intonation, even during World War II! Remember Hitler's craze?"

"That's all I need to think of now, Vera!"

"Ok, if you listen to old recordings of politicians' speeches from those times, you will find that their intonation had a much higher amplitude and sounded like they were preachers in one of those off-the-beaten-track churches. If we listened to them now, they would sound deranged and out of control. I don't think anyone they would make it past a few minutes of public speaking, speaking that way."

"Not all were like that."

"But do you see a pattern? Overall, the politicians of modern times are much more restrained in their intonation, and so they are less of brain-launderers than those back in the day. The restraint in their thought parallels the restraint in their idea expression.

Just like 'uhm', intonation is highly contagious. What is most terrifying is that 'catching' the intonation of another comes with absorbing the emotion, the mental setup, and the worldview that this intonation carries. The old generation of politicians' intonation was more effective at spreading its mental virus, by virtue of its being more aggressive."

"So, how you speak affects how you think? It is like, when you make yourself smile, you actually start feeling better. Interesting."

"Yes, if you start speaking like queen of England, you just might start thinking like a queen—eventually!"

"Where does intonation come from?"

"From the people around you, obviously, but each individual has their own intonation as well. How much you are susceptible to the intonation you hear around you depends on your emotional immunity—strength of character. The easier it is for you to catch a new intonation, the more open your mind is emotionally to the world—and the easier it is for you to learn a foreign language.

There is one intonation I noticed lately, in the 2010's: it is slurring the ends of words, lazy drag through consonants. In its algorithm, it carries alienation, insecurity, and depression. Just please, don't start talking like that!"

"You are scaring me! Is there an intonation that will tune me for happiness?"

"Yes, of course! Seek out the most empathetic and cheerful people you know, and listen to them. Record them and have their speech play on the background, as you are studying, as you are driving, as you are walking—and that will restructure your brain to make you happy. Intonation is an auditory algorithm that makes you who you are."

"Oh là là! Hold on then. La-la-la…"—started humming Elsa. It was 'Vocalise' by Rochmaninov, one of the most beautiful melodies of all history.

"Ha! Somehow, this explains the power of a religious incantation, and of all singing: it all makes sense now. You become what you sing. You are what you hear…"

THE BASICS OF THINKING

"Finally!"—I said with my eyes closed, cheerfully savoring the image of what I was just about to say.

"Vera, wake up! I must have lulled you to sleep with my singing. We talked about everything, and almost nothing on Artificial Intelligence. I am ready for some straight and square answers. Consider this a formal interview."

"Hold on, Elsa! I just saw the most beautiful intonation ever. Let it sink in and tone my brain a little."

"Again? Vera, if you have nothing new to say, then just say nothing."

"What am I now, a news reporter? Ha ha! You know what I realized, Elsa? No matter how much I tell you, I will never tell you everything I have to say. And all of it is new. Speaking of straight and square answers, I will tell you straight and squarely—I don't like to package my thoughts in straight and square boxes, because that is exactly what nothing new feels like. No, I mis-spoke. That is often how the unknown looks like to the prejudiced, and since it is the unknown, it might not even be anything new."

"If you are building me a mental labyrinth, then…"—Elsa started saying, when I interrupted her.

"I am not building anything, I am just letting you into my labyrinth. I guess, you just don't really like it."

"Vera, if this is your idea of helping me understand you, then…"—Elsa protested, and once again I interrupted her.

"Why am I writing this book, Elsa? One of the infinity of reasons is because I hope to help science in understanding how the human brain works, so that it could be reproduced artificially. I am at all times implicitly implying, punning, alluding, hinting, and referring to Artificial Intelligence, even if I don't say it straight and square. The truth is that I need help making sense of what is happening in my head. I can only describe what I see, but its interpretation is something else entirely."

"Well, give me something."

"If only I could think of how to think straight and squarely, I would give you anything you want, Elsa. But then, you wouldn't have any emotional context for the zany ideas that are still to come."

"There are more zany ideas?"

On Artificial Intelligence

"I am almost ready to divulge! Let me explain. What I really want to understand is how this relatively tiny organ can do work that is millions, if not billions, of times more complex than any computer we currently know, and with so little energy. It must be a completely different way of functioning! What I see in my Phantom Reality looks so much like the Artificial Intelligence diagrams I have seen in the textbooks! The networks and their nodes, the systems and meta-systems of networks, the human inability to see how an algorithm got its answer—they are all identical to what I experience inside my brain, where these processes are visualized.

The first time I saw them, I felt like an alien ship finally came and shone a light on me! The diagrams and illustrations I saw on the pages of AI and some math textbooks, even without fully understanding them,—I had seen all that before, inside my brain! It was that unbelievable moment of truth when you feel so overwhelmed, you detach from reality, because it is just too intense. I felt like I started breathing through my skin. The tortured curiosity of all these years and the million of inklings that tickled my brain all this time suddenly converged into singularity. I finally understood that all those things I had been seeing in my brain—they were real and they made sense. It felt like I was exonerated of all crime after sitting in prison for a hundred years under false conviction. *Yes! The Artificial Intelligence as we imagine it, it already exists in our brains, we just have to get it out somehow.*"

"So, what you see in your brain matches exactly what we know now in Artificial Intelligence?"

"It looks very similar! That is why I would like to call my book 'Ideasthesia, something, something... Artificial Intelligence in the brain.' A lot of it, I can only describe, but do not fully understand. I would love to have scientists help me interpret it. Elsa, also one more thing before we begin the next section. Whatever I tell you here will never be the full picture of what I experience in my brain."

"You have said that before. What matters is that you tell me something new. Are you ready?"

"I have been ready for years!"

"Where do we start?"

Brain to Reality Feedback Loop

"I will start with the basics. In order for an experience to be understood by the brain, it has to be converted into basic neural structures that can be used by the brain to store the input, to find and match it to similar experiences, and to relate it to other experiences for later retrieval. In fact, in the brain, memories, imagination, what one sees on the screen, and what one experiences in reality—are all processed in the same place—probably, everyone's personal Phantom Reality.

This is a very rough model. Various incoming sensory patterns cause certain chemical reactions that emerge as neural structures in the brain. After being processed, these neural structures cause human actions in response. The human actions have an effect on the environment, whose reaction is fed back to the brain as sensory input, as a chemical punishment or reward, which closes the feedback loop, mechanically teaching the brain what actions are successful and which ones are not. This chain reaction reminds me of a Recurrent Neural Network, unsupervised learning, and back propagation in Artificial Intelligence.

The punishments and rewards are based on the matches or mismatches between the expectation created and the subsequent feedback from the environment. If it is a match, then the pattern of action is reinforced. If it is a mismatch, another trial is launched. The trials are random chemical commands, within a certain range, that cause physical actions. So, if a child misses the mouth by a centimeter, the new trial will be within several centimeter range, and not within a range of meters."

"How is the matching done? What exactly is being compared?"

Mathematics in the Brain

"This is what I see. There are actual physical interconnected structures made of neurons, seated at various points in space. Relative lengths of their connections, the angles between their connections, as well as paths that their connections are all compared. Everything is measured based on the constants in nature, using physical matching and ratios, rather than numbers. One of such constants that I can identify is the ratio of a circle's circumference to its diameter, known as pi. There are others—all reflections of a single underlying constancy. Brain's mathematics is nothing like the necessary evil that many of us associate with our normal math. Humans only use numbers because we are used to counting objects. In the brain, just like in nature, there are no objects, there are no true borders, points, lines, or curves, because everything is infinitely connected."

"How else are we to discuss nature, if not with numbers?"

"Maybe we just need another way of looking at it. In fact, Gödel's incompleteness theorems and Russell's paradox reassure me that, mathematics as we know it, is only descriptive and even tautological. What if the math I see in my Phantom Reality is something that will work?"

"What makes you think it is different from regular math?"

"Maybe it is not so different. Maybe it means looking at the same thing from the opposite direction. For example, why do we calculate pi as a ratio of circumference to diameter of a circle, producing an irrational number, whose digits go on forever? What if pi is a constant, a whole, and a base, in terms of which we should be counting everything instead? Human mathematics is object-based, but mathematics of nature is based on relationships. In the brain, there is no zero!"

"But how do you signify nothingness?"

"There is just no notin' there! There is no nothingness."—I laughed. "The closest thing to a zero that I see in my Phantom Reality is either a point of change of direction, or where there is a perfect match of entities that are compared. Zero is always where the whole becomes greater than the sum of its parts. Everything that happens in the reality depends on these 'zero' matches, especially growth, change, and most importantly—the apparent time."

"This makes me think of common opposites, for example, man and woman and yin and yang. Together, they would be a zero because they are a match that creates something greater than their sum, it produces life… I wonder what drives the brain to do its thinking, to do anything at all."

"Let me draw you some structures of what I see as part of calculations based on pi, in my Phantom Reality. It is all about angles between sides, some of which I call vectors, and they are what is being measured. I am sorry that I cannot manage to say anything more clearly. These structures slip through my vision as they please."

A napkin was probably invented for times such as these: when one is cornered into their own esoteric hole of incomprehensibility, of something that most likely appears to be an outstanding derangement to the outside world. Then, one is obliged to deliver something tangible, by doodling low profile, non-committal, obscure images that justify one's preceding possible nonsense. What a cool trick!

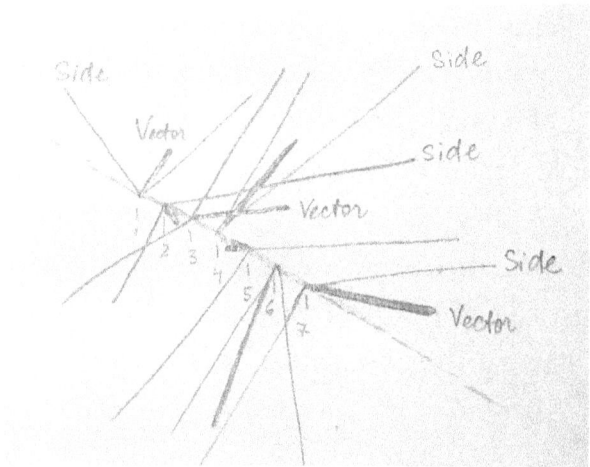

One thing I know for sure: it is quantum computations that come the closest to what I see as the mathematics in my brain, in their use of angles for calculations. That, together with visualizing neural network as a complex adaptive systems, are the most telling analogies I can think of..."

"This looks something like a 'Neural DNA' strand!" —exclaimed Elsa.

"Well, here, these neural DNA strands are made of parts,"—I squiggled a few more things, while Elsa focused on the movement of my fingers, at the expense of processing my drawings. "They are made up of these!"

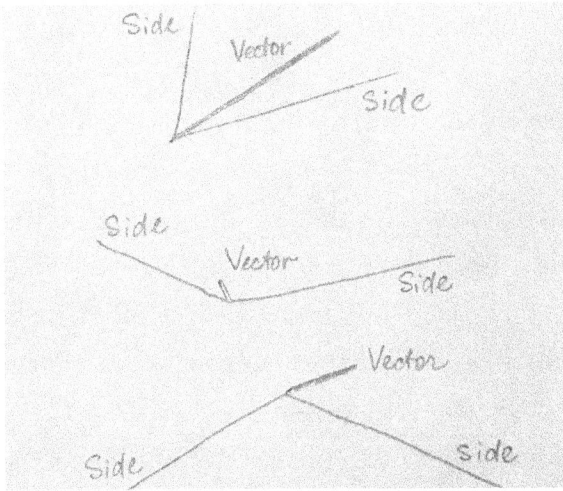

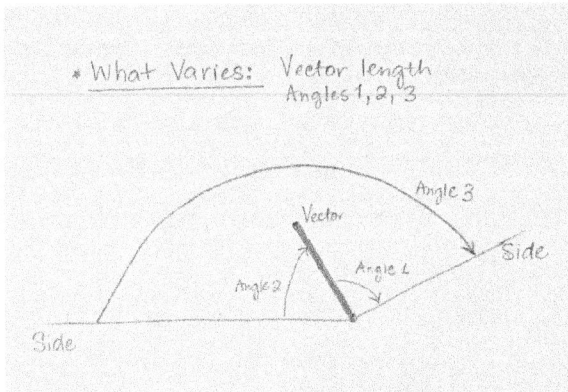

* What Varies: Vector length
  Angles 1, 2, 3

"Vera, I am not sure I understand your doodlings."

"I am not so sure I understand them, either, but someday somebody will!"

Vision in the Brain

I looked up into the sky and took a moment to reflect the sunshine on my face, while my mind was reflecting on what I just had said. A moment later, I looked at Elsa with little sparks of mystery and excitement in my eyes, and said:

"Remember, we talked about all the senses originating in one single channel? Vision is all about navigating the physical space, so as to avoid a falling tree, not get eaten, in order to eat ourselves, and to procreate. Those blobs of life who do it well—survive. Others all died off a long time ago. So, it is vision, along with its four vassal senses, that drives a live organism—I think.

The sensory loop we just talked about—it was all originally created to serve the vision. If you think about it, you will see that it is not perfect at all! Our eyes are not as sharp as those of an eagle, but sharp enough to recognize faces and dangers specific to us. Our bodies don't move through space with the precision of a humming bird, but well enough to avoid obstacles, make love, and have our hands bring food to our mouths. In the environmental niche, onto which we happened to come upon in history of the planet, what we have is just good enough, and not more. How banal... Yet, look at the beauty and complexity of the results!"

"I see, therefore I am..."

"And you know what else? The way the brain recognizes images is the way we recognize star constellations. It is the way we understand a story. It must be also the way we understand tango, because it absolutely takes two."

"And it definitely takes two to finish this book!"

"Alright, alright! I hear the call of the Artificial Intelligence. The difference between how I see the vision in my brain..."—I started saying, when Elsa interrupted me, this time, stretching her arm to my face and covering my mouth with her hand. My mouth completely covered, I tried to mumble another version of what I wanted to say: "Machine Learning has states and actions, but my Phantom Reality tells me that it only has vectors and directions that..." Elsa interrupted me once more, but this time with a smile and, withdrawing her hand from my face: "Ok, this is better. Go on."

"There is just no way to pinpoint a state in the brain, just as it is impossible to pinpoint a point in space: there is always a way to zoom in and see a point as a cluster or a sphere, and it is in constant motion.

Current Machine Learning needs many-many examples for learning, to recognize an image: it studies images bottom-up, from every single pixel to the big picture. Our brain processes the world top-down: it learns a basic pattern, and then it looks for what it already knows. Machine Learning seems to process linearly, even if through layers, but the brain processes images all at once, where the value is in the very interaction of all the parts. Human brain needs only a single time to learn a pattern, and it will use this pattern to search and recognize it in the input that it is given. Human brain is a stereotyper. It is biased to first impressions, and wonderfully primed for the confirmation bias.

When a child is taught what a dog is, for the first time, the brain creates a basic mental structure to represent a concept of a dog. It looks very much like a basic neural structure—with nodes and lines connecting them. This structure can be stretched on a continuum, appearing less and less like the original dog structure, up to a certain point, when it just stops resembling a dog altogether—yet, that point is almost impossible to pinpoint.

Recognizing a dog is the same as instantly recognizing a triangle or a square. Or a letter of the alphabet, or a number, or a hieroglyph, or a face! The basics of a triangle are three connected points, but the angles do not matter. The basics of a square are four points and four equal sides, and four equal angles, where the sides can be stretched proportionately.

In this way, a number three or any letter, or any person's face, or anything at all has stability in our perception: even if it gets stretched or enlarged in scale, or even if it turns toward us at an angle, we still recognize what it is—because of the stability, evident in the preserved relations and ratios between its angles and lengths that our brain is tracking. Once again, it is all about angles, directions, ratios and matches—and all of it is calculated on the basis of the constant pi."

"Interesting."

"That's all? You are like a moomia!"

"What's a moomia?"

"It is the basic structure of your face when you have that reaction."

"Just remember that I am the one who is interviewing you here, so don't chew on your feeding hand."

"Really? I thought, I was the one feeding you ideas!"

"It is a feedback loop between us, Vera!"

"Oh, I see! So, who is the reality and who is the brain here?"

"I am definitely the reality!"—chuckled Elsa.

"I'll gladly settle for the brain. At least, I am sure that I exist!"

"Mademoiselle Brain, I will have to direct the flow of the information here. So, all that vision and feedback loops, and all that snickerty-dook stuff— how does it happen? In other words, how do all these gnarly neural structures communicate? Since you are the brain here, I thought you would know."

"Look who is in a good mood here all of the sudden!"

"Just tell it like it is.   With you— I have to entertain myself.   That's the way it flows here. Otherwise, I would be gone a long time ago. Jus' give it to me!"

Communication in the Brain

"Ok, ok, Madam Reality.   Here, eccolo, na! Communication within the brain happens through something that feels like electricity or a magnetic field. Physically, it looks very much like the waves of color running in the skin of a chameleon. It feels the same as how the muscles get their impulses to move: through electrical surges that are so fast that they are virtually instantaneous.   In the brain, the complex 3D neural structures are simultaneously aware of all parts of themselves.   It is probably because electrical surges continuously run through them, updating the connection.   Any particular combination of neural structures is activated with an electromagnetic rush. Because such 3D multi-layered hierarchies overlap and are interconnected, they are aware of many 'worlds' at the same time: each node of each network is also a member of many other networks simultaneously.

"I wonder, how many degrees of separation there are between any given nodes?"

"Me too."

"Do nodes move around? How do they know which one is which, and where each one is located?"

"They move around, while preserving their connections. Instead of x, y, z spatial coordinates for the location of each node, the nodes are marked by their relationships to other nodes within hierarchies, which are marked by the paths of their connections. It really feels like kinship ties—exactly! It is like defining what a family relative is—let's say an uncle. How do we determine your uncle's coordinates within the larger extended family? Your uncle is either your mother's or father's brother, he is also a sun of so and so, he is also a father of so and so, a husband of so and so, and a cousin of so and so. In turn, all of those people to whom he is related are also related to others through similar kinship ties. That is how everyone in the extended family knows and recognizes each other.

Neurons communicate with each other and behave exactly like a flock of birds, mixed with a beehive, and with an ant colony: everyone has a history and a specialty. The brain is just a huge extended family, a society of neurons who know each other, directly or indirectly, but they all have seen or heard of one another at some point in their lives. They cooperate and work based on their kinship ties. Each brain is different, just like each society is. All, things being equal, I wonder what would happen if we could have two societies meet—how shocked and surprised all those neurons would be! Could they unite and work together? Or would they fight each other?

Brain hierarchies of hierarchies are social networks, and have a memory with whom they have worked and whether it was successful. The reason you cannot build a human brain from parts like LEGO is the same reason why you cannot buy yourself a relative or a friend. It takes time to build relationships, one can never become one's blood sister, and so it takes time to grow a brain from the smallest group of related neurons."

"I have been thinking, are we a network of two?"

"We are a network of one, ha ha ha! Because only one of us makes any sense! Did I read your mind on this one, Elsa?"

"I wonder what your brain does to think such lovely things."

"Happy to please the feeding hand. Oh, since we are at it, I will tell you all about the toolizification of my brain."

"Tool-a-what?"

"Tool-a-brain-drain!"

"Are you in Gagaland again?"

"I just got so excited, thinking about what I am about to divulge. Get ready! You will hear all about my tools of the trade!"

"I am all eyes, and my ears are already flapping!"

MODELING TOOL

"Ok. So. Modeling is not just how the brain creates physical action, but it is also a thinking tool for my personal use in the Phantom Reality. It starts with a question that I launch, throwing all the idea ingredients together, and then watch them transform into a new system, pattern, or a shape. Sometimes, it all organizes itself into a fractal-like chemical dance. Sometimes, nothing but a single entity comes out as an answer, and sometimes it is a dynamic system that can be further mentally manipulated, in order to answer more questions. Sometimes it takes time to appear. I can toss it and turn it all as I please, see all of its parts at the same time, as if in an X-ray machine, and be at any of its points simultaneously. When I examine Modeling in my Phantom Reality, which essentially means modeling the Modeling, I realize that it is the most basic and commonly used tool by everyone. How could I know about anybody else? I really don't—it is my Modeling tool that told me so.

Further, analytical thinking is really creative thinking, because, at least in my Phantom Reality, to analyze something means creating a model of it and running thought experiments in different scenarios—all of which is creative work. I cannot analyze anything without modeling it first. Time, ideas, numbers, any mental constructs—all have physical and chemical qualities there. Asking a question is bringing these qualities together in a Modeling interaction, where their inherent nature will lead to a natural resolution.

The ability of the brain's neural network to model reminds me most of a Boltzmann machine, where every node is connected to all others, where the system is dynamic, unsupervised, and self-determining. Interestingly, its capability is reminiscent of a stem cell cluster—it can model whatever it is exposed to, it becomes the object it studies, and then can know everything about it, without ever contacting it again. It does that by 'throwing' together all parameters, limitations, definitions, and tendencies together, creating either a complex adaptive system, or a stationary outcome. Either way, it now holds a 3D dynamic big picture of whatever it is analyzing, and can predict its behavior by tweaking any single node in it."

"Could you draw something to illustrate what Modeling is?"

"The vision of it in the Phantom Reality is so drastically different from how we see in normal life that I wouldn't even know where to start to do it justice. I need help with that."

"You will find it, don't worry. More tools, please!

"Take a sheet of your paper and turn it over: there might be a surprise on the other side!"

SWARMING TOOL

"One of my very favorite thinking tools is my Army of Ants!"

"Ok, now I am seriously worried."

"You should be. Because they are coming to get ya' if you snooze!"

"Are you threatening me?"

"You bet, I am!"

"Interesting!"—laughed Elsa.

"So, that's a chi-ching!"

"Akh! What's that now?"

"It is my way of saying 'you are welcome to it.'"

"In your Phantom language?"

"Yes, in Phantom."

"Without further ado…"

"Alright, it only looks like an Army of Ants, and there are no real ants involved or hurt in the process. Formally, I call it my Swarming tool. So, this is when I have an idea of what I am trying to understand, and at least, I know its approximate location in my Phantom Reality. I feel that there are zizilions of weightless particles that launch themselves toward this object and surround it completely. These particles have no weight or radius—they get into every nook and cranny of the object, until no part of it is left unexplored.

You could compare it with water that assumes whatever shape it is offered. At the moment when the object is completely surrounded, it feels like a lockdown and a click, and the object appears to me as the negative space where the 'ants' have not gone. It is like a cast of something. The moment of the click reminds me of quorum sensing in the bacterial world, when a certain threshold of bacterial population serves as a tipping point for them to start acting like a single body organism and attack the host. In other words, they don't attack until they feel enough power in their numbers. The tipping point in the Phantom Reality is when the entire surface of the object is covered, resulting in a click. A lockdown is a threshold of confidence about what the object is.

Once I got the object on a lockdown, I start feeling it as part of my body: I feel its weight, velocity, chemical properties, and personality and structure, so I can model its hypothetical moves and interactions, or create a movie out of it for my entertainment. I can shake it, take it apart, melt it, disperse it, spin it out of control, or make it dance along a crazy trajectory I invent for it. It applies equally to abstract and physical objects. Take even a real tree: I can study its roots, foliage, and bark, feel the juices flowing inside it, then uproot it, shoot it into the sky, make its leaves shiver, all to music, and then pop it back into the ground, like nothing happened. With an abstract object, it is the exactly the same!

I can also reproduce an object ad infinitum, which is something like a 3D printing, you could say. Or, have you seen those laser portraits that they sell in the malls? The ones where they scan you 360 degrees, and then laser your 3D image into a glass cube? Any object that has been locked down can be lasered like that, so that I can see it from the inside, and from every direction at once.

The Army of Ants is essentially tactile vision. Well, it is more like figuring out a shape of something, without looking at it—just by touch. Surrounding any object reminds me of gradient descent—you keep filling in in all directions, until it is full. That is why gradient descent is often stochastic—that describes the random movement of the particles that search for the spaces to fill. It is this randomness that underlies the diversification in learning, and explains why diversification is one of the best strategies to follow. Think of when a company wants to determine what pasta sauce their customers will ultimately prefer. What are your chances of finding out, if you provide only two choices? What if you give them a thousand? With a million types of sauces attacking customers' choices, your probability of learning goes way up!

The scariest thing in thinking with the Swarming tool is a perfectly smooth and round object, with no obvious nuke or cranny to get into, in order to see its internal structure. The Army of Ants scrapes it and pushes itself in, in a common effort, and will zoom in to search for ever-finer openings. This is all about how much you know about what you don't yet know. Due to its difficulty, such an object is studied gradually, and in spurts—or cycles of zooming-in onto a small feature and zooming-out to see it incorporated into the big picture. This works for studying abstract concepts and physical objects as well.

"Do you see now why I like my Swarming tool so much?"

"You mean, your tentacles?"

"Yes, my mental tentacles."

"Just call things their names: your nasty Army of Ants!"

"I am sick of you, Elsa!"—I laughed.

"I am so sick of you too, Vera! I want to pull you up into the sky like that tree and shake all your tentacles off of you, then pop you back into this chair —like nothing happened."

"Your sense of humor is only scary on paper, but in real life—it smells grey and comforting like petrichor."

"The shades of your madness are infinite,"—smiled Elsa.

"Like the number of Ants in my Army!"

"I am turning the page!"

ECHOLOCATION TOOL

"My next thinking tool I call Echolocation. This is something like what dolphins and bats do to navigate their space. I use this tool when exploring something completely unknown—searching for the unknown unknowns. This tool is somewhat opposite to the Army of Ants, in that it often explores space from the inside-out. It is about learning the unknown unknown, and creating space. It has the same weightless particles filling out a space with their stochastic, random movements—Brownian motion— that is what it is! Unlike in the Swarming tool, there is no lockdown, but a continuous exploration, based on call and response, where particles shoot out in all directions, each of them separately reaching a limitation, and then coming back with feedback. The aggregate of all the feedback from all the particles constructs a 3D vision of what is out there.

Just like with the Swarming tool, this method is about seeing through touch, and dividing space into positive and negative, and also like filling a cavity with water. Both Swarming and Echolocation remind me of different concepts in arts and sciences: kriging in Geography, asymptote in Math, negative space and stencil in drawing, Gestalt Closure principle in art, and the Black Hole Horizon in Physics.

Kriging is similar to Echolocation only in that it predicts something from knowing only its surroundings, and without observing it directly. The concept of asymptote is similar to Echolocation in its infinite approach to something certain, which can be deduced based solely on the behavior near the asymptote. Negative space and stencil drawing essentially deal with the concept of dividing space into complementary parts: negative and positive, object and its shadow, figure-ground,— all of which use each other as reference for their own definition and existence.

Gestalt Closure principle reminds me of Echolocation because the lines that define a figure are imaginary, yet they outline the figure nevertheless. It is the same with the particles that surround an object, where the outline of an object is only in the contrast between the collection of particles' multitude and the space devoid of them. Black Hole Event Horizon is very similar in its structure to the Echolocation tool: we do not see the Black Hole itself, but only conclude that it is there by virtue of its being outlined by the matter that has still managed to escape its gravitational pool. I don't know of anything like this in Artificial Intelligence, but something analogous could probably be invented—especially because it can already be directly observed and studied in nature.

I often use the Echolocation tool to understand evidence and argumentation in an internal mental debate, as well as a debate between several people: bouncing around the topics, in order to find out the boundaries of what something is not, in order to get a 3D picture of what we know and don't know. Analyzing pro's and con's, and the many sides of an argument, in order to form a big picture, feels like walking through a pitch black cave with many niches—one for each aspect or argument. As the particles bounce off of its walls, the cave reveals itself and grows. Once the 3D print of the arguments is formed, its parts start interacting with each other, which starts feeling like the Modeling tool. Some arguments limit or completely annihilate each other, some support each other and together feel stronger and heavier, and others form clusters to relate to other argument clusters."

"Echolocation is how some people grope their way to the fridge in the middle of the night!"

"Only those who live in caves!"

"That's why I never do it. I live sub divo. I wonder, how would it feel to examine sky with Echolocation?"

"Bliss, if you ask me."

"You know what the most unfair thing is in life? It is the apparent unpredictability of nature, despite all of its myriads of laws we think we know."

"You are right: all those boundaries, dichotomies, and dualities—they are not real, because in reality—everything overlaps and nothing is fully predictable."

"Maybe apparent boundaries are also just a tool, like increased contrast in photos, in order to help us see better."

"Maybe we call unpredictable only that which appears a mystery to us? What science cannot fathom, we call art."

"To me, art is what affects me emotionally, and only if I couldn't explain why. If I can explain it—the art is gone, and gone is the magic."

"And that is why I am bad at explaining! I don't want to let me magic go! Do you understand?"

"Stop explaining then!"

"On the other hand, there is magic at arriving at an explanation. Plus, explaining is never-ever final, and so the magic is infinite. And that—is something worth living for."

"All I am doing right now is trying to stay alive, waiting for the end of your explanation."

"Then, keep turning the pages, Elsa, to speed up the time!"

SCAVENGING TOOL AND LICENSE PLATES

"New page, new life!"—I shouted into Elsa's face.

"You have nothing new to say, I get it."

"Well, that's an exaggeration!"

"You see, it is all about just the right amount of something"

"Exactly: everything in moderation. Like the font: not too big, not too small, so we can read. Like saying: not too much, not too little, so we can understand. Like living: not too long, not too little, to enjoy life. Can one even live for too long? The trick is finding that fine line, the peak at the top of the curve of optimization, to where all roads lead."

"And it doesn't have to be Rome! There is actually a place in Istanbul, where real roads converge, from all around Ottoman Empire."

"At the right place, at the right time! Yes, that's exactly why this is the right time to talk about scavenging, because that is what scavenging is all about, and it is my next tool for thinking!"

"You scavenge for ideas?"

"I would call it hunting, except the ideas are not running away from me, and there is no one to chase. So, a Scavenging tool!"

"It would probably be appropriate to call it good old searching."

"It is more than that, because this tool feels more adventurous, opportunist, erratic, and unpredictable than just plain search. And thank you for not saying the word 'research'—I said myself, but it is the same as calling myself an old fart."

"You hate research that much?"

Traditional Research and Its Pains

"For some reason, a lot of the research feels painful to read. It seem to imprison my thoughts into

167

what feels like a tight trench that wrings me, as if to squeeze out the last of my life juice. Sometimes, my brain shuts down in self-defense."

"Is it that bad?"

"It is worse than I can describe it to you."

"All the time?"

"Why so absolutistic? Ok, virtually all. If it is a short idea, then it does not bother me. What bothers me is a prolonged explication of every step on the way to a conclusion. My brain is trying to complete a puzzle, while what I read is trying to prolong the wait, it seems. It makes me feel like a miner, suffering the drudgery of the sunless stifling underground, impatient to discover a gem. But occasionally, when I find that rare treasure of explosive insight, I feel myself pouring onto the page in excitement, like Dali's liquid clock!"

"Otherwise?"

"Otherwise, my thinking is free to roam my Phantom Reality. It has spoiled me rotten with its vast spaces, and with its magical ability to perform the unimaginable, where answers are crystal clear, self-evident, and quick. There, the thinking relies on the citations as much as seeing an immense blue sky relies on a citation to prove its blueness. When you look into the sky and its depth swallows you whole, do you need to cite someone else to prove that what you see is valid?"

"That's a difficult life to live."

"It is! Because you barely fit into the system of the inter-citational world around."

"I guess, it is in the interface of research papers that I find myself entangled. Ideas, in their pure form —there is barely anything more beautiful which a human can experience."

"So, there are two conditions that would make research pleasurable: a more user-friendly interface, and good, succinct ideas. Right?"

"In a nutshell, yet. Or in an egg shell. Let me tell you, this morass of documentational protocols, of listing tools and methods, listing and reading dry data, history, dates, and names—feels like being force-fed and violated in the most terrible way. It feels like getting paper cuts all over my body, while eating sharp metal bits and pieces, mixed with plastic and old torn up wires. All this busy entropic work, when you are looking for a simple answer, when all you need is a specific idea to fit into the puzzle of thought you are feverishly holding onto at the moment! It feels exactly like making a cake, holding the very top layer in your both hands, then asking someone to smear the cream on the layer underneath, before the layer you are holding comes crumbling down."

"Ha! How so?"

"It's simple! All you need is someone to smear the cream on really fast—you need that action, so you can make your cake. You don't need them to recite the cream's benefits, its ingredients, where it was made, its expiration date, and whether that facility also manufactures school busses and prints books!"

"Oh, I am glad, I am not you!"

"That is why, I don't like doing research in general. When I have a burning question, I would rather model it all in my head: it is eons faster and more intricately revealing. By the way—all my questions are burning! And, sometimes, I would be glad to not be me, but most of the time—it's a piece of soup."

"You mean, pie?"

"Soup!"

"Ah, because you can't divide soup the way you can divide a pie, right?"

"Exactly. You are starting to understand me."

Neural Structures and Clusters

"I don't know if I understand you, Vera, but something in me answers to the very things I don't understand."

"Well, even I don't always understand what I say. In any case, every abstract concept looks like a cluster of basic neural structures, and can be broken down. A basic neural structure looks something like this! A napkin once again came in handy.

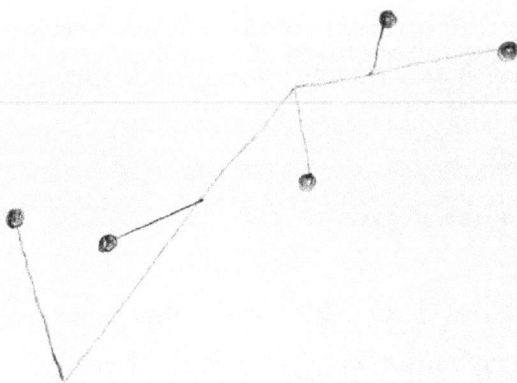

The lengths of the connections and the angles between those connections all carry information, and that is what differentiates all concepts. Such clusters are then further combined into networks that are somewhat dynamic—their nodes move around, or get new connections, or just die. Is that neuroplasticity?"

"It very well may be. Go on."

"Searching for an answer feels like launching one of such wired structures into this dense network of wired clusters—to scavenge around for matches. The structure is looking for structures that match it—in length of connections between the nodes and the angles between those connections. Sometimes, the matches are partial, sometimes—full. Sometimes, it is a whole wired cluster that is sent out to scavenge for its match. When matches are found, they are collected into a string—I call it a License Plate.

License Plates

"You don't really mean that you have a license plate inside your head, do you? Because that would end our conversation right then and there."

"Ah, you never change! Did I even drop the tiniest hint of such an idea? I said, I call it *License Plates*, because the strings which comprise the concepts in the brain remind me of how License Plates have letters and numbers in a row, enabling it to represent a large number of cars. These neural strings of information feel like music chords, ringing together, but they look like License Plates in that they have slots to be filled with one item from a list of options, like numbers and letters for every space. License Plates are the reason why even a very small brain that has relatively few neurons is able to perform complex functions. Through such a system, a short code can represent high complexity, and a small-sized brain can handle vast amounts of convoluted information. If their connections were visualized, you would see that there are hierarchies of hierarchies, overlapping and traversing other hierarchies of hierarchies."

"Ok, that needs a breath or two to imagine."

"Take your time."—I looked at Elsa with empathy. I knew how complicated it must have felt for her to envision things like that, because she was probably using her outer brain, which is not suited well for this task. "Step by step: any basic idea in the brain is represented by a 3D cluster of nodes. I told you about it before. The nodes have no coordinates, their location is in relation to other nodes, where paths and angles between their path are detected, is calculated based on the constant pi. I actually don't know how the calculations happen, but it feels like electromagnetic and chemical reactions that happen automatically, without any deliberate overseeing or directing."

"So, I see the nodes floating in the space of the network. Then what?"

"Every node is part of a neural simple structure, and every simple neural structure is part of a neural cluster, and every neural cluster is part of hierarchy of clusters, and every hierarchy is a part of another hyper-hierarchy. This way, every node has a relation to all other nodes, through different paths, and every node is aware of every other node in the network."

"This sounds like our planet Earth in the Solar System, within our galaxy, within a conglomeration of galaxies, and so on, all the way to the Universe!"— exclaimed Elsa, as if she saw a unicorn landing on our table from the sky.

"Yes, you are right! It starts even deeper—with our atoms, molecules, our bodies, then our planet, and so on..."—I answered, with an equally radiant expression on my face. I almost could see my own glow! "So, now, hold on to this vision of myriads of nodes. A cloud of nodes, let's call it a 3D group, represents a certain concept. These nodes can be very far away from each other, but they are still connected into a structure. Each one of them carries its own information, based on where it is seated, and its proximity and paths of connection to other structures. Don't forget that all structures are paired with sensory stimuli, so they all represent something in the real world. Are you holding it all together now?"

"Barely. Go on before I lose it!"

"Ok, so the concept that a particular group of nodes represents is something that can be literally collapsed into a string of neural clusters, a row of them, just like License Plates."

"How?"

"By stringing all the nodes of the structure into a sequence like a License Plate. A regular License Plate is a sequence with a certain number of slots, each of them being either a number or a letter. For numbers, there are 10 choices: from 0 to 9, and for letters—however many there are in the alphabet. A License Plate in the Phantom reality is a string of slots, where every slot has a choice of many-many possibilities to choose from. That is how a relatively short string can produce billions and billions of possible combinations, all of which correspond to something in the physical reality.

However, it is very different from a simple License Plate sequence, because the choices for each slot are not from a rather short list of letters and numbers, but are pulled from within a category with an enormously large number of choices. Each slot stands for a different category. How the structures are sorted into categories? I can only see that it is based on their shapes, patterns in their lengths and angles, but I fail to hold a more voluminous picture to gather more detail."

"Could you give an example that relates to our reality? Love, for example."

"Funny you would ask that! The concept of love would probably have a slot for the word itself, a few for the aspects of the feeling, a few for memories, one or two for the dreams, a few for the image of the love of your life…"

"I can barely stand hearing about love in such terms."

"That is exactly why I hesitated. Love is so much more than just a string of structures. It is better left not understood."

"Right, or the magic might slip away!"

"The magic will never slip away. Why? Because love can only be felt, and never understood."

"Just like our soul—if we have it."

"We do have a soul, but it is not part of the Phantom Reality or the brain, although we can feel its existence through there."

"Alright, now, give me a funny example. You have the most gawkish sense of humor."

"I do? This is not funny, but a neural License Plate is something like a sequence C5жI, where C is a letter of the English alphabet, number 5 is a number, ж is a letter in the Russian alphabet, and I is a Roman numeral. They are in a string together, but each of them means something in the category and context from which it comes. For example, 5 is fifth after 1, sixth after 0, and eleventh after -5. C is the third letter of the English alphabet, if listing the letters from A to Z. However, what if our sequence were the following: C—my brother—5–love—sky—(an image)—sound of rain—year 1991–memory of a dream—the front right tire of the car of my friend's girlfriend's mother?

The number of possibilities of different combinations is larger than all of the particles in the universe! Something like this is happening in the brain with lightening speed, where decisions are made based on proximity, match, thresholds, regression, echolocation, and associative scavenging. And more! Sounds impossible at first, but each of these methods works wonders in the brain. I can't describe them all, because they fly by so quickly, but I just know that they are there. In short, final answers fetched with a Scavenging tool appear in the form of such License Plates, such strings."

"I think, I understand now..."

"If you expand a License Plate string back to a 3D neural cloud—there is a flash of light!"

"I need a daydreaming break!"

Attention Bias

"Stay with me! What types of neural structures are launched as questions, as well as the trajectory of how such structures travel through the network—describes how a person thinks: what the person looks for, what the person ignores, and the type of logical connection a person makes. I call this Attentional Bias. It is what the person has picked out from the reality, and is then used to build their entire reference system for what reality is and what they are in this reality. We all know this as a bubble!

When asking themselves a question, people simply launch a neural structure they have created to navigate the network they have built for themselves, and it collects the matches, and strings them into an answer. The matches are actual physical structures. This is how we detect analogies between concepts. The focus necessary for this kind of a search has a price—it is a deliberate discard of what does not fit, an intense prejudice and xenophobia. It is like always looking for the color red, and then basing one's whole worldview on it, even if it only represents 1 percent of everything."

"You know what that means?"

"That we are blind to most of what is happening around us."

"It means that our mind is a true Schrodinger's cat, my dear!"

"'Said Elsa, as she finished reading 'Alice in Wonderland', dreamingly turning the last page,'—I mocked Elsa. "I am serious here!"

"Me too."

"Going back to what I was saying. We are breathlessly stuck in the tautology of the dense forests of nodes in our heads. Actually, this is how people turn into optimists or pessimists: they seek out particular aspects of a situation, to the exclusion of others. It is not seeing the big picture, but focusing on a few patterns, while in denial of all others."

"Confirmation bias. I kinda knew that before. But can we ever see the whole picture, and do we ever get out of our mental constructs?"

"We have to learn to tolerate contradictions and cognitive dissonance, to get to what explains it all. It is all about picking out the right piece of soup."

"You mean, pie?"

"That's exactly it, you can't cut soup the way you could have your piece of pie and eat it too. Meaning, you cannot have your mental guards up and learn at the same time. You have to be open to transformation and search it out, even if it changes who you think you are.

Searching for Answers

Some people experience paradigm shifts, and feel like they are walking on water. For some others— sitting in an armchair is drowning in one inch of soup."

"Scavenging through soup, what a primordial idea!"

"Back to the search for answers in the brain. The movement trajectory of a scavenging neural structure usually looks like a coil, spiraling through space, in a repetitive cycle of various moves, which often reminds me of the moves in chess that every figure is allowed to take. Like, if a horse would keep going and going: three forward and two to the side, and again, except, it is much more complex with neural structures. Trajectories that appear more erratic and all over the place—those belong to the processes of learning something new and creativity.

Thinking Coils and Curly Hair

To tell you the truth, sometimes, I can see other people's thinking coils. The more linear they are—the more close-minded those people usually turn out to be. These thinking coils look like people's hair: some curly, some straight, some frizzy. If you study the pattern of people's curls—you will see that they are all different."

"Are you trying to say that people with frizzy hair are extra creative?"

"Notice—you said it. On the other hand— there is Einstein!"

"Please stay focused—I need your Attention Bias on Scavenging."

"I am just Scavenging, don't you see? This is my type of focus: I go on tangents to connect things that are seemingly unrelated, my neural cluster is picking them out of the network. I see them as all related, but to you it looks like I lost focus. It is because your clusters do not move the same way as mine!"

"I am who I am, Vera."

"You are who we all are."

"That… that was inappropriate."

"Just let it pass. Anyway, the chess moves, the coils, the ratios, the trajectories that the neural structures collect through the 3D space of our neural network—they remind me so much of all kinds of tessellations in Islamic Art, in Escher, and very much in Generative Art—but not fractals! Those are all my favorites not by chance! Do you know how it feels to see in reality that you only see inside your head? That is a real proof of one's existence!"

"Something else exciting?"

"Yes, discovering something previously unknown, a new combination, a new association—that is what a shower of goose bumps in the brain feels like! There is another thing. I already told you that there is no zero in the brain's math. The only thing that feels closest to a zero in the brain is what feels like an explosion, when a very special type of a match happens. You see, these collections of matches are seen as a curvy pattern, different from the License Plates. It is a pattern of a patterns. Each pattern is some kind of a complex curve, with all kinds of peaks. When two or more of such curves are combined, some peaks coincide, and when they do—there is an explosion of action! This is what I call to be at the right place at the right time, the real zero!"

"Enough excitement. Do you have any problems with anything?"

"Yes, but problems are exciting, too!"

"Tell me about it! I mean, tell me about it."

"Speaking of explosions. This happens very often: I ask myself a question, launch a neural scavenging structure into the boonies of my neural network, and instantly—there is an explosion of ideas in all directions, and I cannot keep track of it all. I quickly lose sight of it, and so does my memory. The rate of this growth feels much more than just exponential. It feels like an infinite growth. The rate of change, and the rate of the rate of change, and so on, ad infinitum—they all seem to increase infinitely fast. When I model it in my brain, I can only feel Big Bang inside me, and I can think of nothing at that moment. It makes me feel intensely alive, but sad at the same time. Sad to know that I have missed all those ideas in that explosive atomic mushroom."

"We all have limitations to keep us sane."

MEMORY AND CRYSTALS

"Vera, do you remember everything you told me here?"

"I try not to. That way, when I think about it again, it will seem new to me."

"Where is memory located in the brain?"

"In my Phantom Reality, same place as imagination. Memory is what holds the entire neural network in the brain. Memory is network, network is memory. It is during sleep that the brain grows them, like webs, really."

"I wonder, maybe the Native Indian dream catcher has anything to do with that?"

"Normally, I would say no, but it is always better to over-notice than fall victim to one's Attentional Bias, if one is out to discover things.

The Nature of Long-term Memory

Memories grow like corals, like trees, like snowflakes, like mushroom societies. All memories are related to all others, just like branches of a tree, through degrees of separation, of course. It is the same as people being related to each other through relationships they have had."

"This reminds me of the stupidest pickup line of all times: 'Have we been married before?'"

"The obvious answer is 'I don't remember!'"—I laughed out loud.

"That always keeps it new!"—smiled Elsa, with her teeth pointing to the sky.

"On a serious note, memories are sequential directional trees of neural structures, and the more complex a pattern is, the more involved and pervasive a memory becomes, and the easier it is to recall, because of its 'tentacles' being present virtually everywhere. Memories look exactly like disfigured snowflakes! They grow like icicles, and their structure is crystalline—exactly like in crystals. Of course, human brain tissues are soft, so in essence, they are soft crystals."

"Liquid crystals?"

"Ok. Sure. Maybe."

"Are all types of memory like that?"

The Nature of Short-term Memory

"Long-term memory is like that. Working memory feels different. It feels like a clump of muscles that strains itself to keep itself together, and I have to keep focusing on it, for it not to relax and dissipate. It is something like a mother's nipple that tightens together to spurt out the milk for the baby. For short-term memory—those nodes in the muscle clump form a relationship, and they will never forget the time they spent together, because they will now be connected as a result. Now, they are more related to each other, and the next time they 'meet', they will have common memories, and it will be easier to work together because memories precipitate and scale, and any future work builds upon what has been done before.

When something slips out of mind, it feels like a muscle clump slowly relaxes, as the memory cluster dissipates into the network space. Each clump is actually a cluster of different multi-hierarchical clusters. It is a cluster of clusters of clusters, with an unfathomable number of possibilities of connections and meanings, and how they come together and connect with each other for just that fleeting moment is the essence of our rumination."

"You sound like Rumi."

"I wish. You know what 'rumi' means in Arabic, don't you?"

"A turkey!"

"So you knew it when you told me that I sounded like Rumi!"

"Of course, I did!"

"I will not forget all about it!"

"Oh, you will! Just rumi on!"

"So, memories appear to have the same neural structure as the questions or searches in the brain. They are the same thing! Remembering something is essentially asking the brain a question, which the Scavenging tool will take around, tracing its way along the memory lines."

"I guess, that is why they say that we are what we remember."

"Even what we see now—it is only what we remember from a moment ago. The present moment... does not really exist. Only memories do!"

"You are going to make me cry now."

"There is no now, Elsa. If you cry, it will be in your past already."

"Even if you imagine the future, it is still in the past!"

"Very often I forget something, and the only thing that remains is a vision of its neural cluster, mixed in with a colorful texture, in my Phantom Reality. The colorful texture is more resilient, and will often remind me of the neural structure associated with it, but then, it all might abruptly stop there. I cannot possibly remember what it all refers to in physical reality! Often, only these colorful vestiges remain, floating around in my Phantom Reality, making me nostalgic for all that I have long forgotten.

Bringing back a memory feels like pressing a thin sheet of paper over some rough textured surface and rolling your finger over it, so as to force the impression into the paper itself. Have you done that with a pencil? The more the paper absorbs the texture —the more the memory emerges from amnesia. It can be very painful. So it is in life: there is the underlying deep texture and there is a thin paper impression."

"What is the easiest thing for you to remember?"

"Ideas."

"What is the hardest thing to remember?"

"Anything dry, like dates or names, unless I look into them closer and see their personalities."

"What is the hardest thing to forget?"

"People. It is not even what I once thought or felt about them, but their stories. The stories that their faces tell me."

Crystals

"How does memorizing actually happen in your Phantom Reality?"

"Back to long-term memory. Neural structures are formed and paired with sensory stimuli coming into the brain. They can be now copied and reproduced, in order to start searches, or in order to write down memories. It is exactly like a written language we use to write our books, except the brain's language is in 3D space, and instead of letters—it has nodes and connections between them.

Information is stored in structures analogous to the structure of crystals. If you ever see frozen water, snowflakes, icicles—those, I find, are the best visual representations of memories in my Phantom Reality. In fact, both water and stones have the same internal structure as memories. In water, the structure is more dynamic, but in stones—it is as if the time there passes slower.

Although reading of the memories is done through neural structure Scavenging, for a scientist to enter the brain and read the information stored in the crystalline structures of memories, there would need to be an equivalent of a Rosetta Stone. The information is *there* in the structures, but trying to extract its meaning it is like trying to read through any of the existing indecipherable scripts. On another thought, what if this could ever shed light on the famous crystal skulls and the information they contain? There is one held by the British museum in London. I wonder what it can tell us…"

"I am glad, you make these tangential connections. It keeps me awake!"

"My mind tends to float, so let me tell you what happened to me just recently! A few days ago, I came into a small grocery store, just looking around. Browsing through the ice cream section, I noticed a huge grotesque icicle cluster on the side of one of the ice cream containers. The icicles were up to 2 inches long, and they looked like a family of hedgehogs anarchically growing its needles in different directions. It was so scary! But then, this vision pulled me somewhere back into a dark area of my brain, teaming with tiny flickering lights, and I realized that these icicles I am afraid of is something I had seen before. This is what memories look like!"

"Outlandish!"

"So, yes, memories are written in a language. It is our brain's language, just sitting there, waiting to be deciphered."

"It is a Gordian knot, I am afraid. We need someone like Alexander the Great and slice it in half!"

"If you slice a mystery in half—you don't solve it, you kill it. If you break Rosetta Stone into pieces— you still have not deciphered it."

"I wonder, if the brain uses a language to function, what does that say about how human languages work?"

"I have something to say about that."

LANGUAGE WITHOUT WORDS

"Vera, do you think that good people also talk about words, or only about ideas?"

"They definitely talk about good words."

"But aren't all words good, and the bad ones are only bad because of what we make them mean inside our own head?"

"Well, if you have Synesthesia, Ideasthesia, and all-kinds-of-sthesia, then it is more complicated. I already told you about that."

"So, tell me something new!"

"Just like all interactions between the brain and its environment, human language is only a tool for matching and repetition."

"You lost me there."

"Think about it, some arbitrary object, word, feeling, sound—anything—is paired with an equally arbitrary thing. They become a pair."

"Ah, it's all about love again…"

"The whole world is about love, Elsa, haven't you noticed?"

"So, they become a pair, for what mad reason?"

"So that, in the absence of one of the lovers, the other will always remember it, so much so it would seem as if they were there."

"Just give me a practical example."

"No, let me tell you, it is just like the Phantom…"

"I know, I am just pretending to not understand. Sorry, just for the sake of practicality of getting things done…"

"Oh, I feel better now… For a second, I thought that, after all, you really were one of those dry-rind people from whom I have to hide my true self."—I sighed in great relief, trusting that Elsa understood me.

"Vera, if you say one more sundry word like that, I will not help you any longer. If you want me around, then you have to stay on track. Grow up and stop being so sensitive, would ya'?"

I could barely suppress the gush of tears that suddenly pressed and burned on my eyes from the inside.

"You are not the one I love, so it is ok,"—I said.

"Are we done weaseling? Good."

This seemed to be the loneliest moment ever. Here I was, pouring my heart out to someone who was caring, on the one hand, but was so mercilessly caustic on the other. Why was she helping me? Was I a pig for letting her do it? Suddenly, I wasn't sure if I could go on with this.

Some say that it is a definition of insanity to continue doing something over and over again, expecting different results. But isn't it also the definition of perseverance? We should not forget the value of repetition and the accumulation of time. You keep climbing the mountain by taking another single step up, over and over again, until one day—you arrive at the summit. You keep digging for water, one shove at a time, over and over again, until it gushes out like my tears were about to, from the inside of my eyes. I think, doing something over and over again is a definition of hope. Sometimes, time is the best action to take. Sometimes, you have to be insane to arrive at victory.

I lost myself in thinking. Elsa was a good person, and despite her sharp edges, I appreciated that she was there for me. After all, we must give a chance its chance.

I decided to continue with the notes, because I could feel hope inflating my lungs, feeding every quantum particle that was ever a part of me.

"Elsa, we started talking about pairs. Pairing some sounds with an object, an idea, or a feeling means that every time you hear this sound—it will remind you of its other half, its referent."

"Like Pavlov's dog, you mean?"

"Yes, Classical Conditioning at work, sure."

"Pairing a word with that sound will make you hear that sound, every time you read that word. And the other way around, when you hear the sound, you could easily imagine how to write it. That is matching and repetition. The basic mechanism of human language, as I see it in my Phantom Reality, is the same feedback loop that the brain uses to interact with the environment. Human language is just another added representational layer, that replaces the action of pointing your finger to a tree. You just say 'tree', and the other person will get an image of a tree. This common shared knowledge of what a tree is—is the reason we can get rid of all physical reality and still experience it through nothing but words, like in a conversation or a book.

Moreover, the brain allows humans to utter a certain word, having already modeled and previewed the result. If you call somebody a bad word, your brain better give you a heads up on what is to follow! How can the words hurt us physically, if they physically don't exist? But they do, and it is all about the pairs we just talked about, the ones that make Pavlov's dog salivate!

In fact, words are chains of physical reactions that we simply never see, so they seem like magic. It all starts with learning a word, which is pairing its sounds with a feeling of pain, experienced before. As when someone calls us a bad word and hits us: at this moment, the sounds of this word become associated with this pain. Sounds are air vibrations that reach our brain, causing a chemical reaction, and thus the pair is consummated. The chemical reaction to the sound is linked with the chemical reaction to the experience of pain. There are so many other ways, too! But, once the pair is established, the presence of one of the partners will automatically conjure up the other: air vibrations—the pain, the pain—will conjure up the sounds of a word. This pair is a physical, chemical entity, providing a word with physical existence.

Imagine, an author writes down some thoughts on a page. After the author dies, even if the last physical evidence of their writing is burned and erased, there will always be that one last reader who remembers what the page said, and what the words meant. And in the brain of the reader, it is still physical, in the form of neural structures! And even after that person passes away, there will still be another link in this chain of endless memory, because every minute transaction in the universe forever stays traceable to its source. The magic of words is in their invisible eternity. I know, because I have seen it in my Phantom Reality."

"That means that any word you say, stays forever, somewhere in the universe…"

"Even any word you think."

"And if you don't think in words?"

"The universe is full of such thoughts. It might explain all the dark matter we cannot account for, until now! I can almost feel them, hovering around."

"That would be great for a horror movie!"

"I agree. But listen to this: humans instinctually search for the remnants of meaning, looking for answers in the sky and in the stars. We even talk to God, with our eyes closed, hoping for a sign of answer, probably searching for it in our Phantom Reality. We grow superstitions, just so that we can communicate. We see a language in everything that breathes, moves, or just is simply there, and even silence is a sign of agreement, as they say."

"Yes."

"Superstitions are a language because it is that same pairing of an event with its phantom partner—the sign. Then, when the sign is there—we wait for the event to occur. Even if it works only a small percentage of the time, we still keep hope. Hope survives against all odds, even if it is only one single glance, one single instance of pairing. Such are humans, despite all odds."

"Did you notice how patient I was?"

"Ha ha! Yes, and thank you. So, I will go on," —I said, hiding my imaginary tail.

"Keep it practical."

"To the brain, there is essentially no difference between a pattern in the music it hears, the clouds in the sky, the aroma of roses, or in the way Chinese is structured—those are all languages, to be dealt with in the same way. Even riding a bicycle is communicating! It is in how fast we ride, where we turn, how we stop, and how steady we are when we ride. You can tell a lot about somebody, just by watching them ride their bicycle: their physical fitness, possible inebriation, their emotional state, and how sloppy they are."

"Yes, and don't forget the intonation, the gestures, the handwriting, eating and fighting!"

"Yes, yes, we talked about it. Language is unavoidable, it is everywhere."

"So, is there a Universal Grammar for which everybody has an instinct, independent of their sensory experiences?"

Universal Platform for Grammar

"No, not a grammar of words and sentences, but a deeper layer that has to do with the primary human goal—navigating the physical world. This layer contains objects, actions, and qualities of all kinds. A person, an animal, a stone, water and air, as well as brain, mind, conversation, love and time—any pattern is considered to be an object. Any change in the pattern is considered an action.

Objects and actions all have qualities, which depend on them, like a shadow depends on the object casting it. Qualities are nothing less than existential parasites."

"That's harsh."

"Well, without an object or an action, there would be nothing to describe. A human language is equally descriptive, and it wouldn't even exist, if there were nothing to communicate. This way, the grammar of each of the world's languages is nothing more than a collection of sounds that pair up with objects, actions, or qualities, in different ways. Each language reshuffles its own syntax and morphology, as if pulled from a general pool of linguistic possibilities. No matter what style of grammar is chosen, it is just another way to express the same underlying physical reality, in reference to the universal human experience in the physical world. So, the universality is not the grammar that all humans are born with, but in the common physical experience they all have.

By the way, in mathematical terms, any word or sentence, is nothing but a string of variables. And variables vary! Like in 'fore-see' and 'over-see', you can swap prefixes to change the meaning, and you can say 'I go to school' or 'I go to work'—and swap the words. That's the basics of it.

If you see a boy holding a book, you could say that the boy is holding it, or that the book is resting in boy's hand, for example. To walk over a bridge, you could say it just like that in English, but in Russian, you would literally say 'over-walk over bridge.' It even makes sense in English! The pool, from which each language chooses its grammar might have certain limits, but they might be determined by how much a human brain can handle. For example, a word that is a thousand syllables long would not be readable.

"How about abstract things? How do languages refer to them grammatically?"

"It is the same way as they refer to any physical entity, because all abstract thinking is conceptualized in physical terms in the brain. I think, it is in agreement with the idea that all humans have Synesthesia or Ideasthesia!"—I exclaimed, elated. "We even have figurative speech for that, when we talk of feelings as if they were physical objects, as in 'love is fire'!"

"Here we go, back to love again. Could you tell me more about the mechanics of it all?"

"See, how you said 'back to love', as if love was an object in some location? Different languages use different parts of speech to represent objects, actions, and qualities. So, in English, objects most often correspond to nouns and pronouns, actions—to verbs, and descriptions—to adjectives, adverbs, prepositions and different particles. For example, 'beautiful' is an obvious description of an object, while 'quickly'—is an adverb, a description of an action, such as in "run quickly." Even an interjection "eh!" is a description."

"Of what?"

"It would depend on the context. For example, in 'You are cold, eh?'—it is a description of an action of 'being cold', in terms of truthfulness of the statement. It could also refer to the feeling that the speaker feels, when saying it, which again would be description of an action, or even of an object. In sum: whereas objects, actions, and their qualities are universal to all human speakers of all languages, the parts of speech, morphology (how words are formed), syntax (how sentences are formed), pragmatics (how language is used) and semantics (what words mean)— are all language-specific."

"Wait, how about something like 'if'?"

"Even 'if' is a description."

"Of what?"

"Of guilt."

"Ah ha ha! Whose guilt?"

"The one who comes first. The effect thinks that the cause is guilty."

"I would think that it is more of a description of space and time. If your time runs from left to right, then whatever 'if' refers to—will be to the left of what happens if…"

"In any case—human language is only an interface for communication, and not a thinking tool. In fact, translating my thoughts into the format of a language is my second favorite sport after thinking! Language is a very poor reflection of what really goes on in a human brain, but it works well enough for us to survive and go on in life."

Sapir-Whorf Hypothesis

"Haven't you heard of the Sapir-Whorf Hypothesis? Clearly, the language we speak influences the way we think."

"Only marginally, and only at particular times. So, when we first learn a sound of a word—it wouldn't do a thing to change how we think. It is when we learn what it means in the emotional, physical and cultural context that we are influenced. With time, the value of the meaning deteriorates. That is how we get cliches! Then, we are reminded of the meaning in certain circumstances that reinforce what the word means. As when someone tells us that we are a good person, and they are obviously happy. In this case, our vicarious feeling of happiness is paired with the word again, and the meaning is reinforced. If we learn a particularly rare vocabulary word and never use it again—we forget it, sometimes to the point where we would not recognize it if we hear it or see it written.

So, the string of sounds that comprise a word is only effective when it is sufficiently reinforced by the environment. Ultimately, the environment is always stronger."

"I understand and I agree."

"Agreement always sounds like a bright rainbow!"

"Does it smell like peaches?"

INTUITIVE PHYSICS

"Let there be light!"—I said, looking up into the sky with a big smile.

"In the days of yore, we would have said Fiat lux!"

"You and your sporadic Latin. What a party-trouper!"

"Ha! Never heard that before about myself. Sounds like an ant eater."

"To me, that word has the taste and the smell of someone who steps on my ant hill—putting it euphemistically. My Army of Ants is out to get you."

"I am only a simple pattern in energy. And I mean no harm,"—smiled Elsa with her face wide open. "Well, enough with niceties, let's get back to the notes, before the sun sets, and it is time for me to go."

"What a Cinderella you are!"—I said straight into Elsa's face, and then mumbled to myself. "Everything is a pattern in energy, and I did not read this in a book, I saw it with my own eyes—in my Phantom Reality."

"What is more basic, energy or information?"— Elsa asked me suddenly.

"Unexpected, but exactly what I wanted to talk about. They are one and the same."

Elsa looked away and stayed quiet, staring into nothingness, for five good minutes, then turned back and looked at me as if she had an epiphany to share.

"Why are you saying nothing, Elsa?

"I have been saying nothing to you for a good half-an-hour!"

"Yes, that's exactly it!"

"Time is relative in the hands of others. If that was a thirty minutes of nothingness, it felt like a century! Well, time might be relative, but there have to be some constants in this world, to prevent zaners like you from exaggerating beyond the imaginable. You are a fun pattern in energy too, Vera— in that you are like a pendulum with irregular cycles of intolerable, irritating irrationality, alternating with intolerable nothingness"—Elsa spitted out.

"What's gotten into you?"

"I am tired of everything being relative to everything else."

"At least, energy and information are constant, in a sense that they are the same."

"Tell me more, and I might take my insulting compliment back."

"Anything to make that happen! It's funny, but I haven't studied Physics enough to understand myself what I will be telling you now."

"Well, that's encouraging!"

"What I mean is that I don't know how to relate what I see in my Phantom Reality to the physics I know. I must confess that I deliberately read little on this topic, so as not to prime myself into searching out the concepts I already know, and ignoring everything that seems out of the ordinary."

"Everything in your Phantom Reality seems strange! Everything. And, I am sure glad you know that you know nothing."

"I am glad you associated me with Socrates, if only for a split second."

"I only meant that your convoluted thinking was almost ok with me. Go on."

Constants in the Phantom Reality

"What I see in my Phantom Reality is that there are various constants which explain everything, and here are three of them: pi, which is the ratio of a circle's circumference to its diameter, the speed of light, and the Sommerfeld's constant—which is the strength of electromagnetic interactions. All these constants are all interrelated, and all refer to something like a degree to which a certain coil is tightened, or how much the universal balance is disturbed—that is how I feel it. If these constants were tweaked just a little less or more, our whole universe as we know it would never come to exist.

"We probably would not have angles of a triangle add up to 360 degrees, and turning around would not mean turning 180 degrees, and straight angles would probably be anything but 90 degrees."

"Yes, it is all about the degree of how much something is bent in this world, some universal setting that determines everything we experience.   These constants indicate the speed of time and the curvature of space, and how strong gravity is. Imagine a clock with arrows. The speed of the arrow closer to the center is always slower than it is at the tip. The difference between those speeds is also constant, somehow related to pi—but why? This in nature is like a federal interest rate, around which the whole economy of the country revolves around. It's like that. Or think of the dollar as the referent currency for the rest of the world. Those would be the constants whose setting trickles into every single economic interaction that exists.

There is another thing that these constants effect. It's the type of infinities we have in nature. There are different types of possible infinities: those that approach a certain limit, those that accelerate at different speeds, those that repeat a certain sequence an infinite number of times—among others. The existence of the constants reveals that there is no true infinite infinity that is completely unbridled by anything!

Time

By the way, do you know how I know that time exists?"

"If I knew, I wouldn't exist myself, probably."

"The best evidence of time is in the shape of everything that moves: it is their unidirectionality! Plants, flowers, trees have their foliage and the roots, and grow toward the sun, in one direction. Fish, dogs, humans and even comets have heads and tails—they move toward someone they love."

"Or someone to kills and eat! Or to headbutt—in the case of a comet."

"And the head always goes first."

"Wait, people don't have tails! Oh, no, I forgot you have one!"—laughed Elsa.

"We can always tell right from left—only because there is a front and a back."

"How about water, or air, or the sand that blows across the Sahara desert? Do they, too, have a face?"

"Each invisible droplet of water has a face, and it can see where it is flowing. Every molecule of air does too. Every grain of sand flying across the desert sees the earth, the sky, and the others who are flying alongside it—not knowing where and why. But all of them coming like that together—there is a sense of purpose, and it is ahead, in front of them. And so, there is time."

"That's moving. How about our awareness of time, what is all that about?"

The Apparent Speed of Time

"What I see is that we register time like an old film camera: in frames of awareness. Our eyes take pictures. The more aware we are, the more frames per second we register, the less aware we are—the fewer frames per second it becomes. In times of danger, we become hyperaware, and that is why everything seems in slow motion—there are more frames per second than we are used to. When we are in the state of the flow, or ecstatically happy—we are less aware, sinking deep into our unconsciousness, our awareness of the outside world reassured with only an occasional snippet of it. So, the time seems to go by faster, because we register it in far fewer frames.

What I find most fascinating is that I keep seeing that it is the expansion of space that is experienced by humans as time. It violently expands in all directions, in all points, at a specific rate that relates to the constants we were talking about. Of course, the expansion could be at an infinite rate, it could be slower than infinite, or it could be slowing down, or transpire at a certain sequence of changing rates. That I am not sure about, I can only see the possibilities. With the expansion of space, everything we know expands as well, but we simply do not notice, because everything relative to everything stays in the same proportion. It seems as if, as everything expands, it is also collapsing at the same time."

"Did I hear 'relative' again?"

"You might have imagined it. Please, write it down nevertheless."

"Oh, I will, doubt me not."

Continuum of Size Scales

"I will tell you more. Sometimes, when I entertain certain ideas or objects in my Phantom Reality, I see them very close to me. If they distance themselves, it is only by virtue of becoming much smaller or much bigger! When they approach me

215

again, it is by going back to a size that matches mine. It is the most curious sensation! I also had dreams like that, too.

There is even more. Sorry, but I see size in general as an entirely relative. I actually see a spectrum of different size scales in the universe. As I zoom in and out infinitely, I go through different layers of them, each layer following its own laws of physics! We, humans, have our own layer, where the units with which we operate are objects we perceive as solid. As I zoom in, I see atoms with huge spaces between them, and even more—I see some kind of revolving massless knotted coils! There are no objects there. As I zoom out onto the scale of our solar system, it all starts looking like atoms again. Further out—and there are no objects again.

Looking at humans from the atomic scale, we find the distances between atoms to be so great that we easily imagine how humans could pass through each other unobstructed! That could be, if the atoms did not 'hold hands' to keep a human solid and put together. If they were all on the loose, then humans would be exactly like clouds and passing through each other! Humans already are walking clouds, but there are too many droplets and they are too far away from our eyes, relative to the distances between them, to see each one separately. Plus, they 'hold hands.'

Did you know, we really never touch anything at the atomic layer? We only seem to touch at the human layer because we are too far away and too large to see the gaps between our body and the objects we seem to touch. At the human layer, that distance appears so minuscule that we consider it nonexistent.

The Knots of Energy

What's even more fascinating is that, if I zoom all the way in or all the way out, I start seeing knots! It appears to me that all the matter is made of interweaved knots that cannot be untied, but only followed and understood. Because, if they were untied, everything would explode. This reminds me of the basic neural structures, the vectors, nodes, and angles that make them up, that are used by the brain as 'utensils' for thinking.

There is actually an ancient language called 'quipu' in South America practiced in the Inca empire, based on knots! No one has been able to decipher it yet. I saw and touched one artifact myself—a string with knotted strings hanging off of it—at a museum in Rio de Janeiro! The moment I saw it, the resemblance between its knots and the vector structures I keep seeing in my Phantom Reality earthquaked my brain!"

"No earthquakes here, please."

"Never mind! It gets even more interesting from here!

What is Light?

"Light appears to be a result of a selective disturbance of space and time that imposes a certain pattern of energy only on the areas it affects. It is this selective disturbance of space and time that releases energy and manifests itself as light. Particles and waves are an observable result of such a disturbance. The more intense the light is, the greater effect it has on time, and the greater the damage is to our physical world. If we want to see how our reality really works, we need to look at music under a microscope—because it is sort of like music. Energy disturbance that causes visible light looks exactly like music chords ringing together!"

"That was equally interesting, but no more. Sorry!"

"Ok, then maybe this will pique your curiosity: crystalline light structures!"

"Could you play piano if a piano is not there?"

"What are you talking about? Ok, sure, it is simple: just imagine it and play it."

"Nothing in the world could shake this sturdy skull of mine. I am not here to be curious."

"Oh, then how can the world go on? You are curious, Elsa, you are just in denial."

"As most people are. I really don't know why I do that."

"You don't have to know, you just have to feel why."

"I won't tell you what I feel, because I don't want to know it myself, and you don't need to know."

"I already know that. It is ok, just help me with the notes."

"Then keep telling me your stories, Sheherezade of 1001 split moments!"

"Thank goodness, Elsa!"

"I did not mean it in a good way. It was only meant to show you how tired I was of listening to you."

"You know, I really don't want to know all about you, after all. It is good enough that I see the good that you are doing, without the why of it all."

"Hey, ignorance is bliss!"

"No! Ignorance is how good intentions result in evil actions!"

"Stop lecturing me! I hate indoctrination with passion."

"Passion plus ignorance is a lethal combination!"

"I am as calm as I could be."

"Being exposed to ignorance is as disgusting as being sober and sitting next to a drunk!"

"I am still calm."

"Ignorance is fear!"

"Only of the unknown."

"Really?"

"Got ya'! That was fun."

"O-kay. Are you ready to hear about crystalline light structures?"

"Always ready, as a soldier that just got confused about the directions of where to shoot."

"Just don't shoot me, and we will stay friends!"

"Ok, so, structures? Lights? Crystals?"

Crystalline Light Structures

"It is crystalline light structures."

"I don't understand."

"If you understood, you would get bored with it. Nevertheless. When I have to explain something to someone and they don't understand, it feels like pressing a tray full of gooey material on someone's teeth, in order to take a dental impression. The more I press, the tighter the gaps are filled, and the better someone understands."

"That I understand!"

"Then, let's up the ante! Crystals and coils! Crystals and coils is all I see when I look around."

"You must be hallucinating."

"And knots, made of crystals and coils! It's not a hallucination, I just did an eye operation— switched my eye lens to transitions—micro to telescopic. That's all! Ok, just kidding. Don't be so serious. No, actually be serious! Because this is serious stuff! Being serious a little is like knowing a little—it is very dangerous. It is just enough to screw up everything, but not enough to put it all back together."

"Crystals and coils, crystals and coils! Back on topic!"

"Ha ha! So, yes. When I model the energy that makes up our reality, I keep coming up with a crystalline structure made up of infinite light sources— meaning, there are no spaces between them. The light rays seem to 'wrinkle' and 'fold' around its sources like coils that turn into knots, making the space 'fit in' far more than 360 degrees! It is as if you can experience space and time from many different 'folds', and even the relation of space to time varies accordingly.

Seen from another 'fold', we would not even appear to exist. This crystalline structure of light projects angles everywhere, but these angles are made of coils. Within coils, the notions of angles, lengths, parallels or even points do not exist, because everything curves and coils everywhere."

"So, it is curves and coils, after all!"—said Elsa decisively and stabbed her papers with the pencil. "I wonder: with all the coils spiraling around and all the crystals sparkling everywhere, how is it that life springs out of this soup?"

Life

"Life is where entropy finds its home. When I model it in my Phantom Reality, I see these blobs of energy that, by virtue of their natural movement, get stuck together into systems of interdependence that do not let them escape. Think of velcro! If you put a hook and a loop together, they naturally form a life! Their energies are directed onto each other, and that is what forms a relationship. This is what creates life."

"Feels like a trap!"

"No, it is the opposite of chaos, and chaos is the real trap! It is a total lack of freedom and total destruction, and it is no wonder that chaos needs much more energy to support itself than life. Chaos is not natural to nature. But entropy is!"

Entropy

"That made me feel like a square root of minus one!"—exclaimed Elsa.

"Irrational?"—I chuckled.

"No, lost and imaginary."

"You obviously don't know your oranges from apples."

"At least, I am not imaginary. And I am not irrational. And I am quite alive. Plus I know my tomatoes from potatoes, don't doubt that."

"Would an irrational life still be considered a life at all?"

"What an irrational question!"

"You are right, it is not about bad answers, it is all about bad questions."

"Even bad questions sometimes beget good answers."

"Only on the island of Serendip, that is. I do know that your favorite sport is thinking, Vera,—but exercises in deliberate irrational attacks? That's something new."

"New is good!"

"Not if this new is pure entropy. Never mind! My ears are flapping, my eyes are jiggling. Whatever you want.   Could you tell me something about entropy?"

"Sometimes, I find it hard expressing what I know,   because I cannot fully pull it out from the depths of my inner and into the clear surface of my outer brain.   It feels like a terrible muscle strain, extricating this unpredictable shape of thought out of some mirky and sticky jelly. It is like seeing a cute little icicle, floating on the sea surface, grabbing it playfully, and only then realizing that the entire weight of another gigantic form attached to it and is hiding underneath, as your whole torso is pulled down into the dark abyss.

I will tell you this.   Entropy versus life can be understood through baby ducks and baby chicks. Chicklets follow their mother erratically, they often stay behind and get lost.   That's entropy.   Ducklings, however, follow their mother very closely.   That's life. This is the simplest physical analogy to how I experience entropy."

"And the universe?"

"It is like a giant pulsating heart, with maximum entropy at its maximum expansion, when the heart is compressed and the blood is released, and minimal entropy—when the universe is collapsed, that is, when the heart is relaxed and the blood rushes in. Do you see the opposing movements of contraction and expansion of the heart and the blood's movements? This is what keeps our reality stable."

"Any final words?"

"Is there already a reason for nostalgia?"

"It's just that every word you say is final for that particular moment."

"Then, yes. If we are to understand both our brain and the universe, we really ought to study complex adaptive systems and how they interact through the laws of thermodynamics and quantum mechanics."

"Thank you. Please, say no more. I am tired."

THE END OF THE BEGINNING

"I cannot remember a thing I told you,"—I said to Elsa, laughing.

"Neither do I. And I didn't hear even half of what you said. But, what you said is all here,"—she slapped her hands on her papers. "I wouldn't rely on those, though. Sometimes they lie, you know. Plus, I might not care enough to give them to you."

It suddenly felt as if my brain swallowed my eyes! I tried to control my fear, which was seething by the second, as I held on to the last of my faith that what Elsa just said wasn't true.

"But you promised! Ah, we are so different, Elsa! How do we ever get along?"—I sighed in despair.

"Exactly! You are so much shorter and chubbier than I am,"—laughed Elsa.

"In whose reality?"

"I really don't know anymore…"

"I do! It is *your* reality, Elsa, *your* vision of the world!"

"So?

"So, you might be wrong. And your vision might mean nothing."

"So could yours."

"But you know what's great? It's that they have equal rights to the truth, and you never know which one of them is the one."

"And if you don't know, then what do you do?"

"Then, you go with the one that disturbs you more, the way that energy disturbs reality to produce light."

"Fiat lux!"—Elsa flipped her hands into the sky.

"Elsa! Don't scare me like that again. We have been sitting here, talking for a year now."

"How is that possible, if the sun hasn't even set yet?"

"Remember my specialty?"

"Ah, the impossible."

"Elsa, it seems impossible, but I have been looking so deep into myself that I have seen passed the atoms, to where there is no longer me. All that is left is this overwhelming force, this fundamental disturbance that pulls everything together."

"Love?"

"Yes."

"So, love is that mysterious constant that created all others..? Don't worry about your book, the air remembers everything. I think, it is time now. I need to get some sleep. Keep these!"—and she handed me her notes.

"Thank you, Elsa!"—I said with great relief.

As Elsa was leaving, I looked at her back, and thought to myself: "Good bye, my imaginary friend. Come back again."

"I heard that!"—Elsa turned around and looked at me playfully, as if she expected this to happen. "You know I am you, right?"

"I thought, that was the only secret I kept from you."

"If you need anything else, you know where to find me,"—and she disappeared into the crowd of people walking by.

I looked onto the notes: they were blank. My eyes rolled up to the sky and I smiled.

REFERENCES

Csikszentmihalyi, M. (2016). Flow and the foundations of positive psychology: The collected works of Mihaly Csikszentmihalyi. Dordrecht: Springer.

Hadamard, Jacques. The Mathematicians Mind: The Psychology of Invention in the Mathematical Field. University Press, 1996.

Nikolić, D. (2009, April). Is synaesthesia actually ideaesthesia? An inquiry into the nature of the phenomenon. In Proceedings of the Third International Congress on Synaesthesia, Science & Art (pp. 26-29).

www.ingramcontent.com/pod-product-compliance
Lightning Source LLC
Chambersburg PA
CBHW021140090426
42740CB00008B/860